Science

Exclusively endorsed by OCR for AS GCE Science

Heinemann is an imprint of Pearson Education Limited, a company incorporated in England and Wales, having its registered office at Edinburgh Gate, Harlow, Essex, CM20 2JE. Registered company number: 872828

www.heinemann.co.uk

Heinemann is a registered trademark of Pearson Education Limited

First published 2008

12 11 10 09 08
10 9 8 7 6 5 4 3 2 1

British Library Cataloguing in Publication Data is available from the British Library on request.

ISBN 978 0 435691 79 0

Edited by Tony Clappison
Designed by Wearset Ltd, Boldon, Tyne and Wear
Project managed and typeset by Wearset Ltd, Boldon, Tyne and Wear
Illustrated by Wearset Ltd, Boldon, Tyne and Wear
Cover photo © NASA/Science Photo Library
Printed in the UK by Ashford Colour Press Ltd

Acknowledgements
The author would like to thank the following for their invaluable help in the writing and preparation of this book:
David Brodie, Judy Griffiths, Marcus Rutland, Debbie Veel and Chris Whitworth

The authors and publisher would like to thank the following for permission to reproduce photographs:

p7 Andrew Lambert Photography/Science Photo Library; **p14** Cumulus; **p15 L** European Space Agency/Science Photo Library; **p15 M** Geospace/Science Photo Library; **p15 R** NASA/JPL/Science Photo Library; **p16 L** European Space Agency/Science Photo Library; **p16 M** Geospace/Science Photo Library; **p16 R** Geospace/Science Photo Library; **p17** European Space Agency/Science Photo Library; **p20** Cumulus; **p31** Cumulus; **p34 T** GSFC/NASA; **p34 B** GSFC/NASA; **p39** Hugh Spencer/Photo Researchers/Photolibrary; **p49** European Space Agency/Science Photo Library; **p63** BAS/Science Photo Library; **p76** Cumulus; **p77 T** Susumu Nishinaga/Science Photo Library; **p77 B** Cumulus; **p79** Cumulus; **p84** NASA; **p94** Pasieka/Science Photo Library; **p96 L** Equinox Graphics/Science Photo Library; **p96 R** Dr Tim Evans/Science Photo Library; **p106** Cumulus; **P107** Cumulus; **p109 L** Equinox Graphics/Science Photo Library; **p109 R** Dr Tim Evans/Science Photo Library; **p114** Cumulus; **p115** Cumulus; **p115** Cumulus; **p125** Cumulus; **p126** Cumulus; **p127** EFDA-JET; **p131** Cumulus; **p136** Jonathan Ling/Istockphoto

The authors and publisher would like to thank the following for permission to use copyright material:

p8 Fig 1: [Part 1] Image from the Vpython program 07_spectrum.py by Bruce Sherwood [Part 2] Ron Kurtus/School for Champions; **p8 Fig 2:** Reprinted with permission from *Encyclopædia Britannica*, © 2006 by Encyclopædia Britannica, Inc.; **p59 Fig 3:** Adapted from illustration by Jack Cook, WHOI; **p62 Fig 1:** [Part 1] J. Hansen, M. Sato, R. Ruedy, K. Lo, D.W. Lea and M. Medina-Elizade, 2006: Global temperature change. *Proc. Natl. Acad. Sci.*, 103, 14288–14239 [Part 2] Michael Pidwirny/The Encyclopedia of Earth; **p63 Fig 2:** *Climate Change 2001: Working Group I: The Scientific Basis*, Intergovernmental Panel on Climate Change, Figure 1b; **p64 Fig 1:** Robert. A. Rohde/Global Warming Art; **p65 Fig 2:** Robert A. Rohde/Global Warming Art; **p86 Fig 2:** P.J. Linstrom and W.G. Mallard, eds, NIST Chemistry WebBook, NIST Standard Reference Database Number 69, June 2005, National Institute of Standards and Technology, Gaithersburg MD, 20899 (http://webbook.nist.gov); **p97 Fig 3:** Salters-Nuffield Advanced Biology

Some of the activities on pages 4, 10, 20, 30, 37, 40, 48, 50, 79 and 81 are based on ideas and resources developed by the University of York Science Education Group, originally published for the 'Science in the Environment' Project. These are published with the kind permission of University of York Science Education Group.

We also thank OCR for permission to publish examination questions. Each has been acknowledged in the text.

Every effort has been made to contact copyright holders of material reproduced in this book. Any omissions will be rectified in subsequent printings if notice is given to the publisher.

Websites
There are links to relevant websites in this book. In order to ensure that the links are up-to-date, that the links work, and that the sites are not inadvertently linked to sites that could be considered offensive, we have made the links available on the Heinemann website at www.heinemann.co.uk/hotlinks. When you access the site, the express code is 1790P.

OCR Science

Exclusively endorsed by OCR for AS GCE Science

David Goodfellow

www.heinemann.co.uk

✓ Free online support
✓ Useful weblinks
✓ 24 hour online ordering

01865 888080

In Exclusive Partnership

Contents

Introduction vi

Unit 1 (G641)

Module 1 Remote sensing and the natural environment 2

1 Light and electromagnetic radiation 4
2 The wave behaviour of light 6
3 Light and colour 8
4 Absorption of light and energy transfer 10
5 Information transfer and the hazards of electromagnetic radiation 12
6 Remote sensing 14
7 Analysing remote sensed images (I) 16
8 Analysing remote sensed images (II) 18
9 Limitations of images 20

Summary and practice questions 22

Module 2 Stable and vulnerable ecosystems 24

1 Autotrophs and photosynthesis 26
2 Energy transfer and active transport 28
3 Respiration and energy transfer 30
4 Ecosystems 32
5 Biodiversity 34
6 Natural selection and the development of new species 36
7 The nitrogen cycle 38
8 Nutrient fluxes and feedback 40
9 Fertiliser manufacture and eutrophication 42

Summary and practice questions 44

Unit 2 (G642)

Module 1 Weather, climate and climate change 46

1 Atmospheric circulation and pressure 48
2 Pressure and kinetic theory 50
3 Movement of air in the atmosphere 52
4 Water and covalent bonding 54
5 The polar water molecule and its unusual properties 56
6 Ocean circulation 58
7 The thermohaline circulation 60
8 Evidence for climate change 62
9 Models of climate change 64
10 Analysing climate data 66

Summary and practice questions 68

Module 2 Chemical processes in the atmosphere 70

1 Acids 72
2 Chemical equations and the formation of acid rain 74
3 Clearly a problem 76
4 Oxidation and neutralisation 78
5 Solutions to acid deposition 80
6 Ozone and catalysts 82
7 The effect of CFCs on the ozone layer 84
8 The greenhouse effect 86
9 Greenhouse gases 88

Summary and practice questions 90

Module 3 Proteins and genetic engineering 92

1 Proteins 94
2 Enzymes 96
3 DNA and genes 98
4 DNA replication and transcription 100
5 The genetic code and protein synthesis 102
6 Genetic engineering 104
7 Concerns about genetically modified crops 106

Summary and practice questions 108

Module 4 Options for energy generation 110

1 Burning fuels 112
2 Fossil fuels 114
3 The story of the atom 116
4 The nucleus 118
5 Radioactivity 120
6 Nuclear processes 122
7 Options for future energy generation (I) 124
8 Options for future energy generation (II) 126

9 Electricity transmission and distribution 128
10 Electrical circuits 130
11 Electrical and magnetic fields 132
12 Alternating fields and epidemiology 134
13 Analysing epidemiological data 136

Summary and practice questions 138

Practice exam questions 140

Answers 144

Index 146

Introduction

How to use this book

This textbook supports the OCR AS level in Science. Because the course itself is unique, so too is this textbook!

The science in this course ...

... is all related to the big issues which face the planet in the twenty-first century. Among the issues which this course will cover are:

- the vulnerability of ecosystems to change
- the factors which affect our climate
- the ways in which environmental problems have been tackled in the past and the challenges we face in tackling the issue of climate change
- the ethical dilemmas caused by new technologies, such as genetically modified crops
- how to decide the strategy for providing the world with energy in a sustainable way
- the way in which remote sensing technology allows us to monitor environmental changes.

Again and again you will find that behind so many of these issues lies the shadow of global climate change – perhaps the biggest challenge in human history.

In each case you will find out more about the science behind each of these issues. In the process you will find that you encounter many of the big ideas of science – from across all the main subject areas of biology, chemistry and physics.

This book ...

... will provide you with the background you need to understand the ideas in the course. Each module of the course is divided into a series of double-page spreads. Each spread is a self-contained section telling a particular part of the scientific story, or showing you how to apply the ideas to the kind of problems you meet. You will find the following features in the spreads:

- **Key definitions**: important words that you need to understand – each definition is only one or two sentences long and could be used when an exam question asks you to explain a term.
- **Examiner tips**: suggestions from an OCR examiner about the way in which questions may be asked on a given topic – and telling you what you *won't* need to know as well as what you will. Common misconceptions and particularly tricky issues may be highlighted here as well.
- **Worked examples**: showing you how to solve problems or carry out calculations step by step.
- **Applications**: particular examples of how science is applied – in most cases you won't need to know the details of a particular application but you may find it useful in illustrating or enhancing your understanding of a topic.
- **Activities**: suggestions for further research and questions to ponder – or even short pieces of practical work which you may be able to carry out.
- **Questions**: there are between two and five short questions at the end of each spread which allow you to test your knowledge quickly; at the end of each module there are some longer exam-style questions and at the end of each of the two units there are some past questions which have appeared in past OCR Science papers – remember though that because this is a new course you may not find that past questions cover all the topics you will be studying.

Going further ...

This book will provide you with the background you need to understand the AS Science course. It may not answer all your questions, nor can it hope to cover all the ways in which the science you learn about has been applied. Using the Internet is an obvious way of taking your understanding further – use a search engine to hunt for sites that feature some of the key words you find in each spread.

Links to the most important sites are avaliable on the Heinemann website **www.heinemann.co.uk/hotlinks**. Using express code 1790P.

Many of the activities will suggest ways of searching for further information. To keep right up to date with current events, start keeping an eye out for science and environmental topics which feature in the media – you will find that your science background is invaluable in helping you to understand the implications of such news stories. Above all, use it to help you to decide what you think the truth is about some of the controversial issues which face the world in the coming century.

- **How Science Works**: throughout the book you will find an emphasis on the way in which science and scientists work in today's society. In many of the spreads you will encounter specific aspects of the subjects which provide excellent examples of How Science Works. These examples could be used by examiners as a way of testing your ability to appreciate, evaluate and analyse the work of scientists and the scientific community. You will find a box in many of the spreads to highlight these aspects of the subject.

General exam advice

Your AS Science course will be assessed by:

- two written papers – Unit 1, which is divided into three modules, and Unit 2, which is made up of four modules
- a series of practical activities
- a case study in which you will research a scientific development.

This book is mostly designed to prepare you for the written papers.

How to prepare for exams

A good place to start is with the summary 'mind-maps' at the end of each module. You should try to expand these into helpful revision aids, adding extra detail to remind you of the key ideas you met during the module. These mind maps will help you to see how the ideas link together, but as you revise more thoroughly you could try techniques such as:

- writing 'flash cards' with key ideas/key definitions or important equations
- going back and using the questions in the spread to re-test yourself
- doing the exam-style questions.

In the exam

Be very clear about what a question is asking you to do – you may see instructions such as:

- **state** – write a short phrase to summarise an idea
- **describe** – describe an idea in more detail
- **explain** – use some scientific ideas to describe the reason why something happens
- **suggest** – this means that you may not necessarily have studied this particular aspect but you should be able to use other, more familiar ideas to help you write an answer.

... as well as more obvious commands such as 'draw' (a diagram), 'calculate' (a number), 'plot' (a graph) etc.

Remember that in an exam only about half the questions will ask you to simply remember scientific ideas. The remaining questions will ask you to do something with the ideas – explain observations, analyse data, do calculations and so on.

Maximising your marks

Always look very carefully at the number of marks available for a question – this will give you a good idea about how much detail to go into – a 5 mark question may well need five distinct points to be made. At the end of an answer, stop and read through what you have written:

- Have you made enough points to gain full marks?
- Have you actually answered the question that was asked?
- Are there any statements that might be too vague to gain credit? For example, often saying 'it reacts' may not be enough – make it clear which substance you are writing about.

It is well worth looking at old mark schemes to see the way in which examiners award marks.

Finally – good luck, and enjoy your exploration of science!

UNIT 1

Module 1 Remote sensing and the natural environment

Introduction

Our understanding of the planet we live on – and indeed other planets – has been revolutionised by the development of remote sensing technologies; images obtained from Earth-orbiting satellites are now in common use in a huge range of scientific areas. Satellite images help scientists to monitor and forecast the weather, observe the small changes in sea surface temperature which may herald much more major climatic changes and find out how changes in land use may threaten the survival of ecosystems such as the rainforest.

To understand how these images are obtained and interpreted you will need to find out more about visible light and the rest of the electromagnetic spectrum. You will see how the production of images involves processes such as scattering and refraction, and how other processes – such as diffraction – limit the quality of these images. There will also be a chance to learn about how electromagnetic radiation transfers energy as well as the information which makes up images.

You will be able to compare the new technology of digital sensors with the way in which your eye works. Finally you will get the opportunity to interpret some satellite images for yourself.

Module contents

1 Light and electromagnetic radiation
2 The wave behaviour of light
3 Light and colour
4 Absorption of light and energy transfer
5 Information transfer and the hazards of electromagnetic radiation
6 Remote sensing
7 Analysing remote sensed images (I)
8 Analysing remote sensed images (II)
9 Limitations of images

How science works

During this module you will cover some of the aspects of How Science Works. In particular, you will be studying material which may be assessed for:

- HSW 5b: Analyse and interpret data to provide evidence.
- HSW 6a: Consider applications and implications of science and appreciate their associated benefits and risks.

Examples of this material include:

- Compare the ways in which different types of radiation may interact with the human body (spread 1.1.5).
- Interpret grey-scale images of the Earth and other objects in space made using electromagnetic radiation (spread 1.1.7).
- Interpret false colour images displaying information about three separate wavelength ranges (spread 1.1.8).

Test yourself

1. Name some types of electromagnetic radiation.
2. Explain what is meant by the *frequency* and the *wavelength* of a wave.
3. Describe what happens when light is **(a)** reflected; **(b)** refracted.
4. White light can be thought of as a mixture of three primary colours. Name these three primary colours.
5. What is the meaning of the word 'digital' when applied to an object, such as a camera, which stores information?
6. 'Satellites are in orbit around the Earth'. What does this mean?

1.1 ① Light and electromagnetic radiation

Field

A region of space in which objects experience forces – for example in an electric field charged objects experience an electric force.

Wavelength

The length of a complete waveform.

It is normally easiest to show it as the distance between two peaks. It has the symbol λ and is normally measured in metres, m.

Frequency

The number of complete wavelengths passing a point in one second. It is measured in units of Hz (hertz).

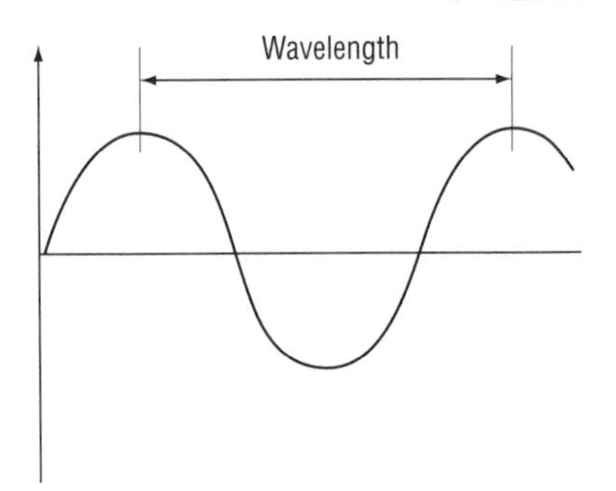

Figure 2 A simple waveform, showing wavelength

Significant figures

The accuracy of all data obtained by scientific experiments is limited by the precision of the method used. When you use data from experiments to calculate a value for a number you may obtain a number on your calculator which has many digits. You cannot possibly be certain of the number to that level of accuracy, which is why you will often find a number written to, say, **3 significant figures**. The number 27863541 shown on your calculator can be written in standard form and to 3 significant figures as 2.79×10^7.

The importance of light

The ultimate source of all the energy used by ecosystems on Earth is the Sun. Most of the energy emitted by the Sun is in the form of what we call visible light. Plants have evolved to be able to absorb and store this energy and animals have evolved the ability to use light to form images which provide them with information about the world.

What is light?

Visible light is a form of electromagnetic radiation.

It consists of vibrating electric and magnetic **fields** (see Unit 2 Module 4). We often describe light as having the properties of a wave. These electromagnetic waves can transfer energy from one point to another, even across the near-vacuum of space.

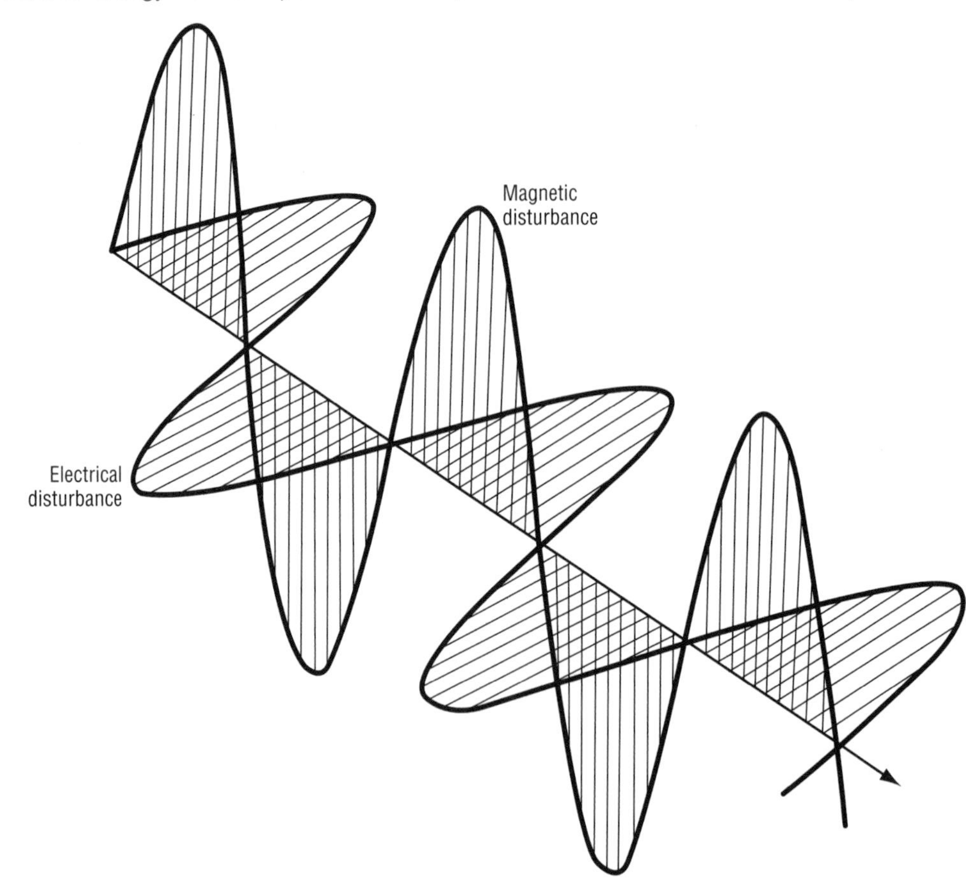

Figure 1 The waveform of electromagnetic radiation

Figure 1 shows the vibrating electric and magnetic fields. Normally it is acceptable just to show a single waveform, as illustrated in the margin.

The speed of light, and all electromagnetic radiation, is $3.0 \times 10^8 \, m\,s^{-1}$.

Standard form and units

3.0×10^8 is an example of number written in standard form – it is written in the form of a number between 1 and 10 multiplied by a power of 10.

The accepted abbreviation for 'metres per second'is $m\,s^{-1}$. You may see it written as 'm/s' but you should become used to the accepted way of writing units. For more information about units see spread 1.1.4 *Absorption of light and energy transfer*.

Worked example: writing numbers in standard form

If your calculator gives a number as 27 600 you may be required to write this number in standard form. Remember that 10 000 is 10^4, so 27 600 = 2.76×10^4.

In a similar way you can write very small numbers in standard form. For example, your calculator may give a number as 0.0000536. You know that 0.00001 is 10^{-5} so 0.0000536 = 5.36×10^{-5}.

Standard form makes it easy to be able to write numbers to a sensible number of significant figures.

Examiner tip

You will normally be given this equation but you may need to rearrange it so that, for example, you can calculate the frequency of a wave if you know its wavelength and speed.

Many people choose to use a triangle to help them remember how to rearrange equations such as this.

Calculating frequencies and wavelengths

Frequency, wavelength and speed are related by the equation:

speed = frequency × wavelength (in symbols $c = f \times \lambda$).

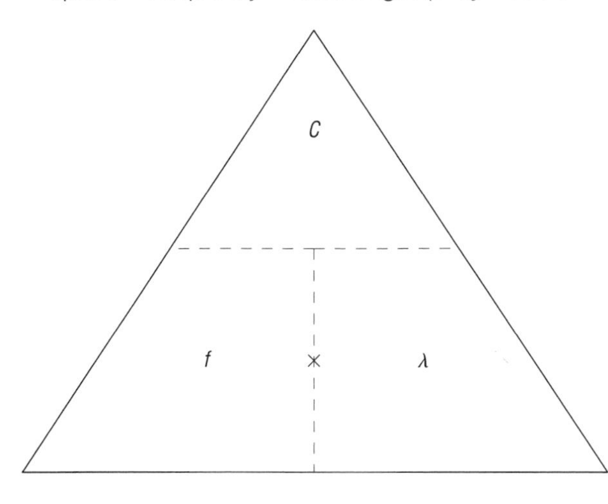

Figure 3 The 'triangle' method of rearranging equations. Cover the quantity which you want on the left-hand side of the equation and the remaining parts of the triangle show you what will appear on the right-hand side

Activity

Human beings have evolved to make use of visible light. We have now developed technologies which allow us to make use of the other parts of the electromagnetic spectrum as well.

Research some of the ways in which these other parts of the spectrum are used.

The electromagnetic spectrum

Visible light is only one form of electromagnetic radiation. The whole range of electromagnetic radiation makes up the electromagnetic spectrum.

Questions

1 Visible light, radio waves, ultraviolet radiation, infrared radiation and gamma radiation are all forms of electromagnetic radiation. Place these types of radiation in order of their wavelength, starting with the shortest.

2 **(a)** Write the following numbers in standard form:
  (i) 9170
  (ii) 0.00169
  (iii) 682 000 000
  (iv) 0.0000000492
  (b) Rewrite these numbers in standard form and to the required number of significant figures (s.f.)
  (i) 28 910 to 3 s.f.
  (ii) 0.01678 to 2 s.f
  (iii) 2 089 391 to 4 s.f
  (iv) 0.000038901 to 3 s.f.

3 The frequency of a type of radiowave is 231 000 Hz.
  (a) Write this number in standard form.
  (b) Use the formula $c = f \times \lambda$ to calculate the wavelength in metres ($c = 3.00 \times 10^8\,\mathrm{m\,s^{-1}}$). Give your answer in standard form.

4 The uses of electromagnetic radiation can be classified as **(a)** using radiation to transfer energy, or **(b)** using radiation to transfer information. Give two examples of each type of use.

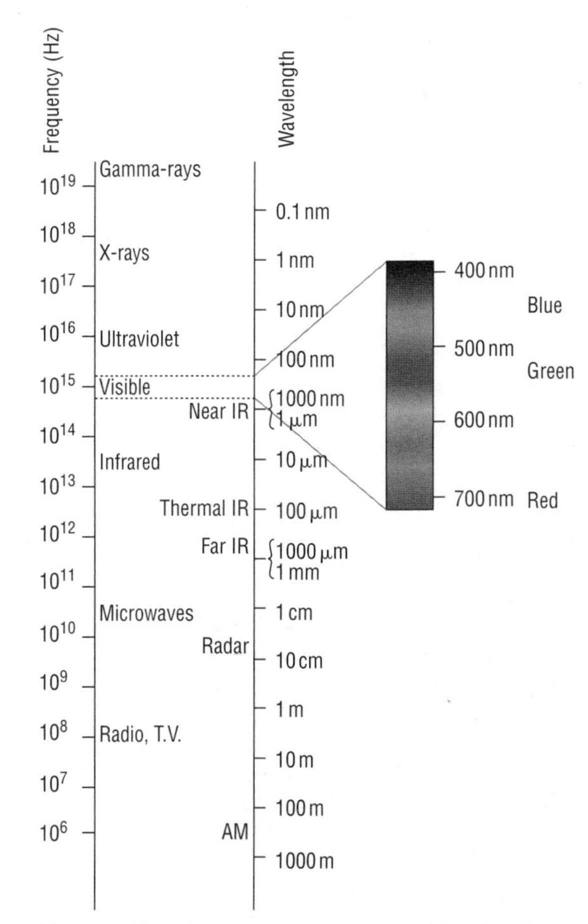

Figure 4 The electromagnetic spectrum, showing the frequencies and wavelengths of the different types of radiation

1.1 ② The wave behaviour of light

The properties of light

You will have observed three important properties of light (and indeed all electromagnetic radiation):

- It can undergo **reflection**
- It can undergo **refraction**
- It can undergo called **diffraction**.

Reflection

A process in which light bounces back off a smooth surface.

Refraction

A process in which the direction of light changes when it crosses a boundary between two surfaces which have different densities.

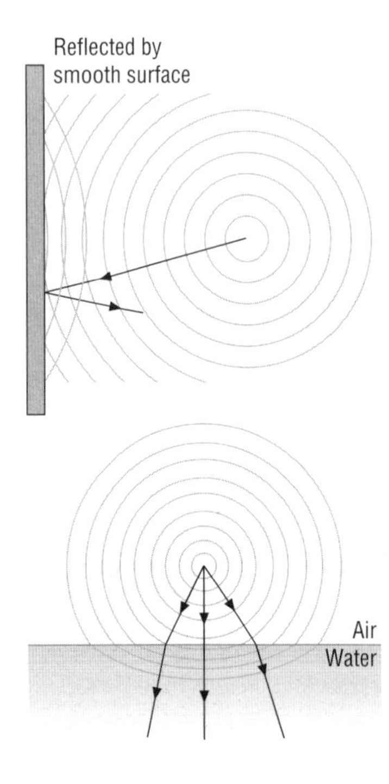

Figure 1 The processes of reflection and refraction using ray diagrams and wave fronts

Figure 1 shows you two ways of showing light in diagrams – either as a narrow ray of light or as a series of 'wavefronts'. The distance between each wavefront indicates the wavelength of the light wave.

These effects can both be easily shown in the laboratory using ray boxes to produce narrow beams of light. However the third property, diffraction, is more difficult to observe and can be explained *only* by using the idea that light is a wave. You may have seen diffraction demonstrated using water waves in a ripple tank or similar piece of equipment.

Diffraction

A process in which waves spread out after they pass an obstacle or a gap in a barrier.

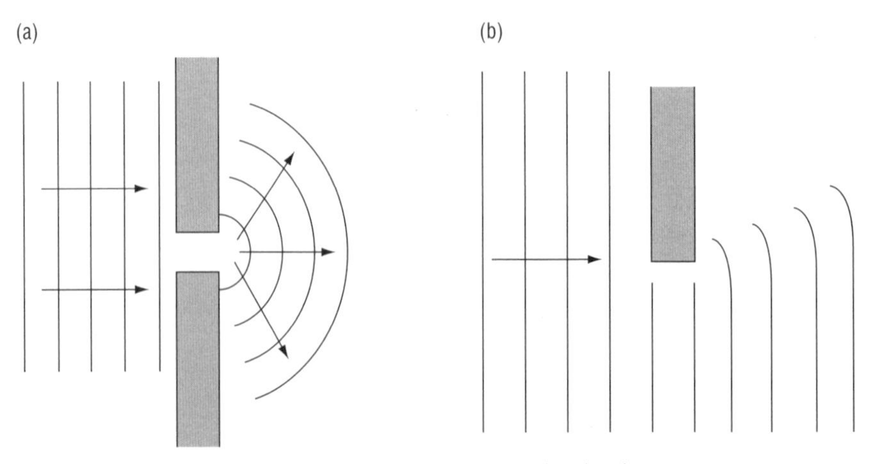

Figure 2 Diffraction occurring as a wave passes: **a** gap in a barrier; **b** the edge of an obstacle

Activity

1. Use a ripple tank or similar piece of equipment to show reflection and refraction occurring.
   Investigate how diffraction is affected by changing the size of the gap in the barrier.
2. A TV remote control is a source of infrared radiation. Devise experiments to test whether infrared radiation undergoes reflection and diffraction.

Difficulties in observing diffraction

Diffraction effects are only noticeable if the wavelength of the waves is similar in size to the dimensions of the gap or the obstructing object. The water waves used in the ripple tanks have a wavelength of a few centimetres (1 cm = 1×10^{-2} m), which is similar to the gaps in the barrier. Light waves have wavelengths of less than 1 μm (1×10^{-6} m). So diffraction will only be observed if the objects or gaps are also very small.

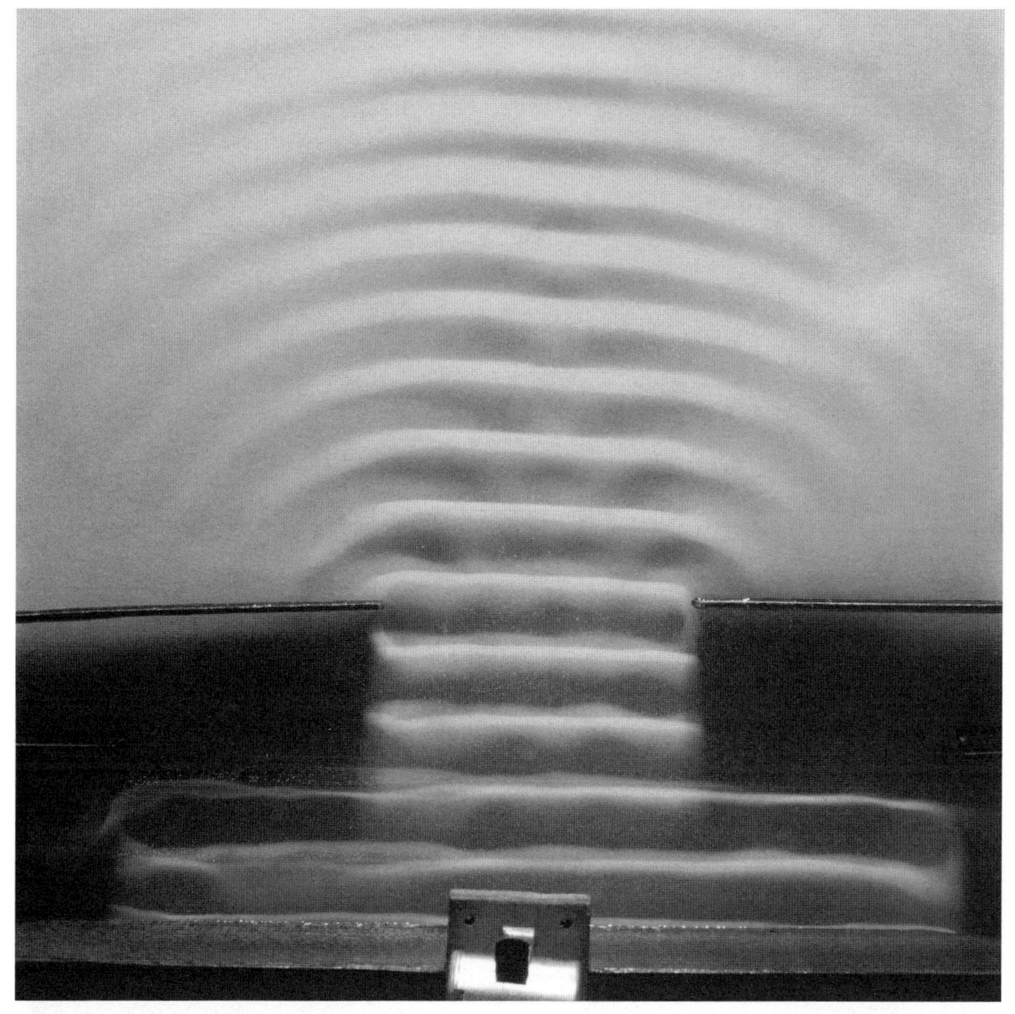

Figure 3 Diffraction effects are hardly noticeable if the wavelength is much smaller than the obstacle

Diffraction effects

Diffraction effects can make observing very small objects impossible using light. For example viruses are typically about 100 nm (10^{-7} m) across. This is smaller than the wavelength of visible light so diffraction will cause significant problems – the blurring due to diffraction will be too great to obtain a clear image.

Diffraction can even cause problems in normal photography. Cameras are designed so that light passes through a narrow gap (or aperture) before it is focused by the lens. If the aperture is small, the diffraction occurring at the edges of the gap will cause noticeable blurring at the edge of the image. This will occur regardless of how good the lens is!

Questions

1 The table below has matched the words reflection, refraction and diffraction to descriptions of the processes incorrectly. Rewrite the table so that the words match the descriptions.

Process	When it occurs	What happens
Reflection	When passing small gaps or obstacles	Light spreads out
Refraction	At smooth surfaces	Light bounces back
Diffraction	When passing from one substance into another	Light changes direction

2 Sound waves have a speed of 330 m s^{-1}. A typical sound wave from a speaking voice might have a frequency of 220 Hz.

(a) Calculate the wavelength of this wave.

(b) Is diffraction likely to be a commonly observed process for these sound waves? Justify your answer – and suggest a way of testing your prediction!

1.1 ③ Light and colour

The white light spectrum

You will be familiar with the fact that glass prisms can split white light up into a spectrum consisting of a range of colours. Each colour represents a specific range of frequencies (and wavelengths) of light, and different frequencies of light will be refracted by different angles.

A diffraction grating (a piece of glass covered with a pattern of lines) can also split up white light to form a spectrum. The lines create gaps of about the same size as a wavelength of light, so diffraction will occur as the light passes through. Again, different frequencies of light will be diffracted to different extents.

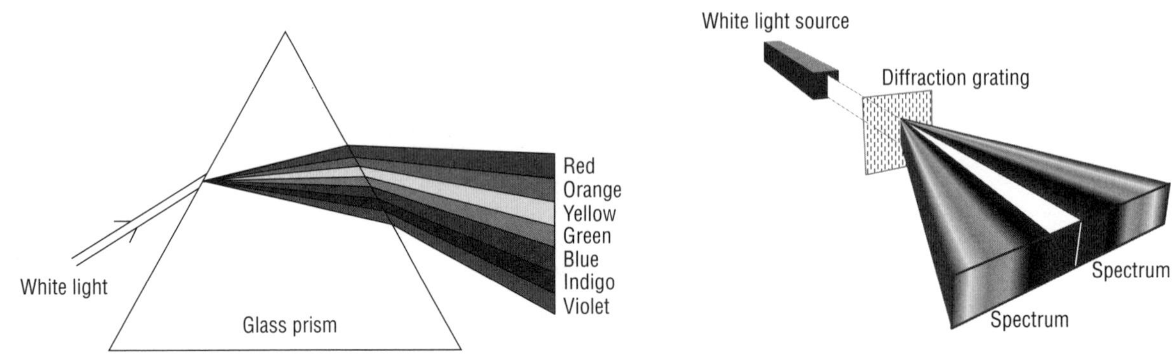

Figure 1 The spectra formed by a prism and by a diffraction grating

The colours of the spectrum

The wavelengths and frequencies of the different colours making up white light are shown in Figure 2.

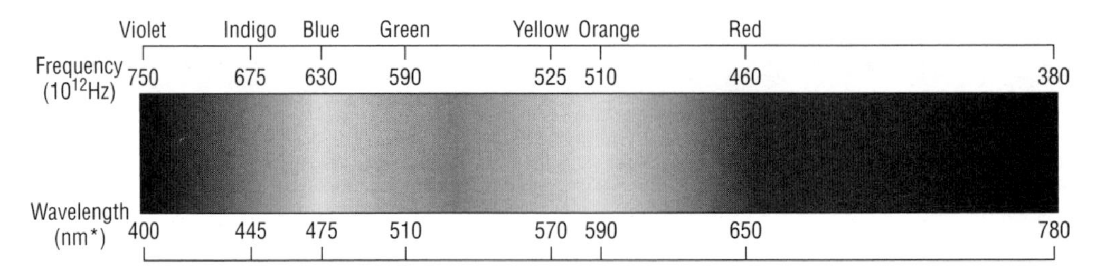

* In nanometres (nm); 1 nm = 1×10^{-9} metre

Figure 2 Wavelength and frequencies of the colours of white light

Seeing light

Not all animals have the same response to electromagnetic radiation. Most animals, including humans, use light to reconstruct the world in their brains, so that they can respond to the world. Different species have different systems, so that they respond to different wavelengths. Bees, for example, can see 'light' that we cannot – we call it ultraviolet radiation. Snakes can detect infrared radiation that is invisible to us – it helps them to locate their warm-blooded prey.

Seeing colour

Although most animals can detect light, not all can distinguish colour. The ability to distinguish different frequency ranges depends on having specific receptor cells that detect only certain ranges of frequency.

Activity

There are several websites which allow you to 'see' flowers as a bee would see them, showing patterns on the flower only visible to bees. Try searching for 'bee ultraviolet'.

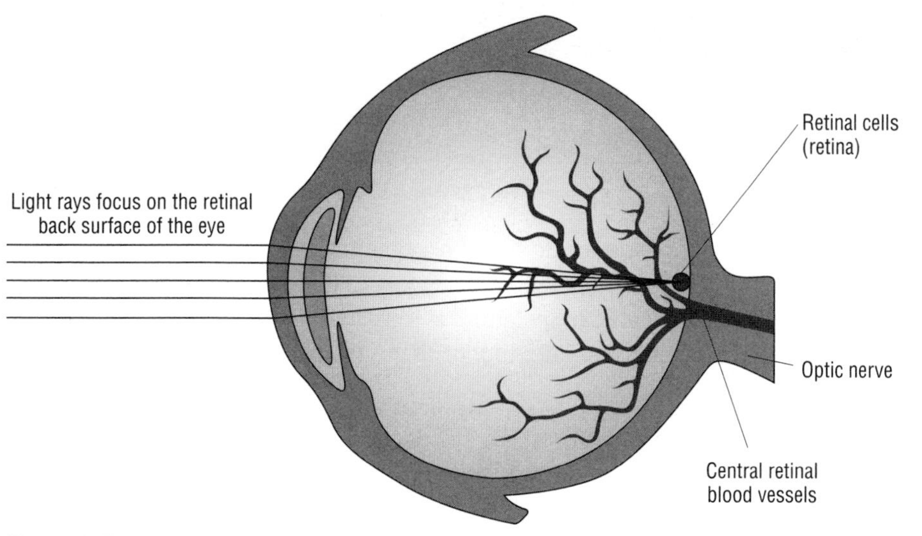

Figure 3 The human eye, showing the lens and retinal cells

The eye has a lens that contains a dense fluid. This refracts light to form a detailed image on the retina. Cells in the retina are photosensitive – they respond to light. These cells can either be rods – which respond to the whole frequency range of visible radiation – or cones, which respond more strongly to particular frequency ranges.

The sensation we describe as 'green' is created when the green cones are stimulated strongly and the blue and red cones only weakly. Similar descriptions can be made for 'blue' and 'red'. All other colours and shades of colours correspond to particular combinations of responses from these cells. For example a sensation of yellow is created when both red and green cells are both strongly stimulated.

You may need to use ideas about the mixing of primary colours when you are interpreting remotely sensed images displayed as coloured images on a computer screen.

Secondary colour	Created from	
Yellow	Red	Green
Magenta	Red	Blue
Cyan	Blue	Green

The key point is that coloured light corresponds to a particular frequency (or wavelength) range in the electromagnetic spectrum. We can recreate the same sensation in our eyes that yellow light produces by mixing red and green lights in suitable proportions.

Remember that these rules for colour mixing apply only when we mix light from light sources. The situation is much more complicated if you mix together pigments with these colours (as you might do when painting). In art, the term 'primary colours' is used rather differently.

Questions

1 Put these colours in order of their frequency, starting with the lowest:

blue, yellow, red, green, violet, orange

2 (a) What colour you would see if red and green light were mixed in equal proportions?
(b) Predict how the colour would change if the proportion of red light was increased.

3 How do you think the response curves of the cones would differ if humans were able to see in the ultraviolet, as bees can?

Applications: response curves of cone cells

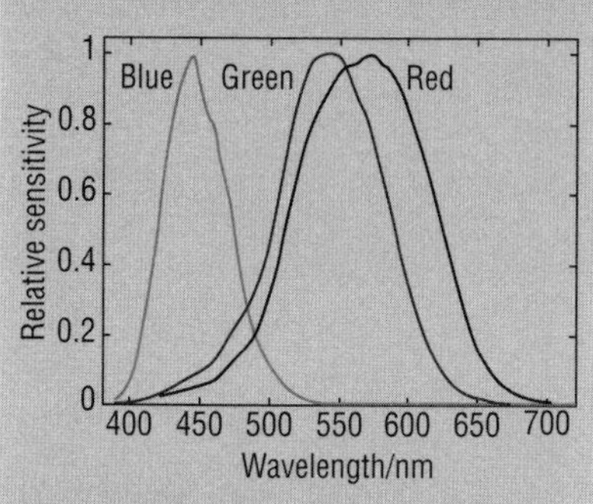

Figure 4 The frequency response curve of the three cone cells

The cones are described as blue, green and red after the colours that fall at the maximum of their sensitivity.

In practice, any coloured light that we observe is likely to be detected to some extent by two, or even all three, of the cone cells.

Primary colours

Colours in which only one type of cone is being stimulated strongly. Mixing blue, green and red light produces white light.

Secondary colours

Colours formed from the mixing of two primary colours (and in which two types of cone are being stimulated strongly).

Activity

There are many colour mixing animations or 'applets' available on the Internet which allow you to see the effect of mixing colours. Find one of these sites and experiment with the mixing process. Many of them represent the intensity of the colour on a digital scale (0–255). See spread 1.1.5 *Information transfer*.

1.1 ④ Absorption of light and energy transfer

Scientists have been searching for life on other planets ever since it was realised that the Earth was not the only planet in the Universe. Some of the search has concentrated on looking for Earth-like planets orbiting stars. The ideal candidate for life would be a planet that has a similar average temperature to that of the Earth.

Why is the Earth warm enough for life?

Some of the light energy from the Sun is absorbed by the surface of the Earth. As a result, the Earth's surface becomes hotter. In fact the Earth then becomes a source of radiation itself and emits infrared radiation back into space. There is a natural balance of energy flow in from the Sun and energy flow back out to space. As a result the temperature of the Earth remains stable at around 290 K.

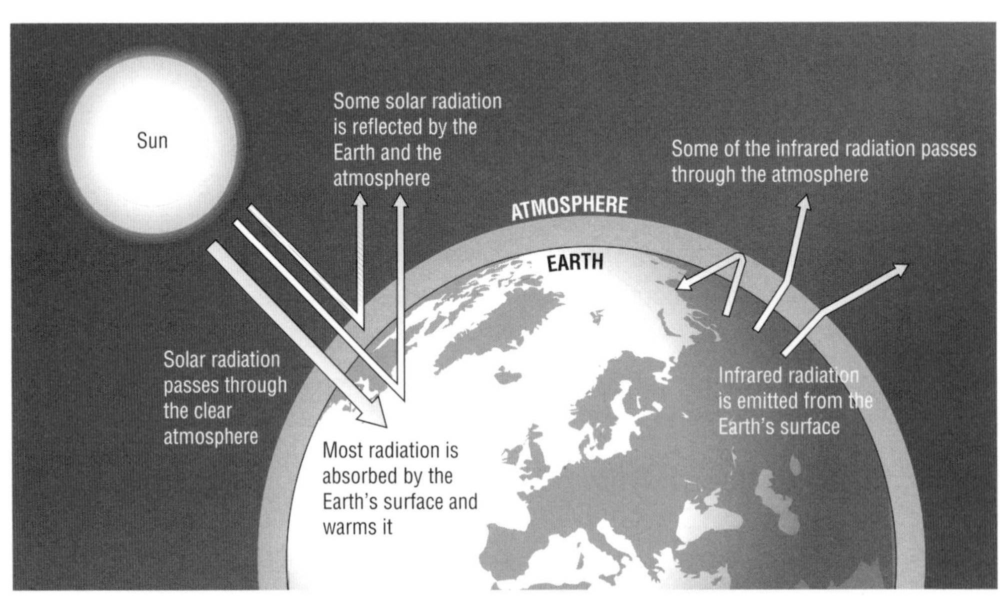

Figure 1 Absorption of light energy by the Earth's surface from the Sun

Absorption by pigments

Any surface that appears anything other than pure white or silvery has the ability to absorb light. White light is a mixture of different wavelengths and some surfaces *absorb* some of these wavelengths more strongly. They *reflect* the other wavelengths more strongly and it is these wavelengths that we can see – making the surface appear to be coloured. Coloured molecules are known as pigments – chlorophyll, the pigment which gives plants their green colour is a good example. Strictly speaking the substance we call chlorophyll is a mixture of two pigments – chlorophyll a and chlorophyll b.

Absorption
400 500 600 700 800
Wavelength/nm

Figure 2 The absorption spectrum of chlorophyll

Activity

You can see the absorbance of substances like chlorophyll very well by passing white light through a test tube containing a solution of chlorophyll (extracted by grinding up a leaf with ethanol and then filtering it). If you then pass the white light through a prism or diffraction grating you will be able to see which colours have been absorbed by the chlorophyll.

Applications: chlorophyll

The structure of chlorophyll molecules is complex but it seems that the pattern of double and single bonds in the molecule allows it to absorb red and blue light very effectively. The light energy is not transferred into heat but into chemical energy, which eventually allows the formation of molecules such as glucose in the process called photosynthesis.

Green light is not absorbed, and so it is reflected – so the chlorophyll pigment appears to be green.

Thermal radiation

The Sun, with a surface temperature of 6000 K, emits energy strongly in the visible region of the electromagnetic spectrum. For much cooler objects, such as the Earth's surface, energy is emitted in the form of infrared radiation. This is often referred to as 'thermal radiation' or 'radiant heat'.

Table 1 shows the typical wavelength and frequency of radiation emitted by the Earth and the Sun.

	Temperature	Wavelength	Frequency
Sun	6000 K	5×10^{-7} m	6×10^{14} Hz
Earth	270–310 K	1×10^{-5} m	3×10^{13} Hz

Table 1 Typical wavelength and frequency of radiation emitted

Radiant heating systems in rooms rely on the fact that hot objects (such as electrically heated filaments or heated ceramics) emit infrared radiation, which is easily absorbed by the surfaces of the room, including people.

The energy balance of human beings

In the UK climate, where the temperature inside or outside is normally no higher than 21 or 22 °C (294–295 K), an unclothed human being (at a temperature of 37 °C or 310 K) emits infrared radiation at a rate of 500 W (500 J s^{-1}). In an enclosed space, such as a room, an average human body will absorb energy from the walls, floors etc. at a rate of 400 W. So, clothes must be worn to reduce the heat loss by 100 W to maintain a balance of absorption and emission.

Prefixes and units

You may find that some frequencies or wavelengths are expressed using units that contain prefixes – e.g. 90.3 kHz or 3.2 μm.

'k' is the prefix 'kilo' and stands for 10^3 (or 1000). So 90.3 kHz = 90.3×10^3 Hz or 90 300 Hz.

'μ' is the prefix 'micro' and stands for 10^{-6} or 0.000 001. So 3.2 μm = 3.2×10^{-6} m or 0.000 003 2 m.

The prefixes you may come across are:

n	μ	m	k	M	G
Nano	Micro	Milli	Kilo	Mega	Giga
10^{-9}	10^{-6}	10^{-3}	10^{3}	10^{6}	10^{9}

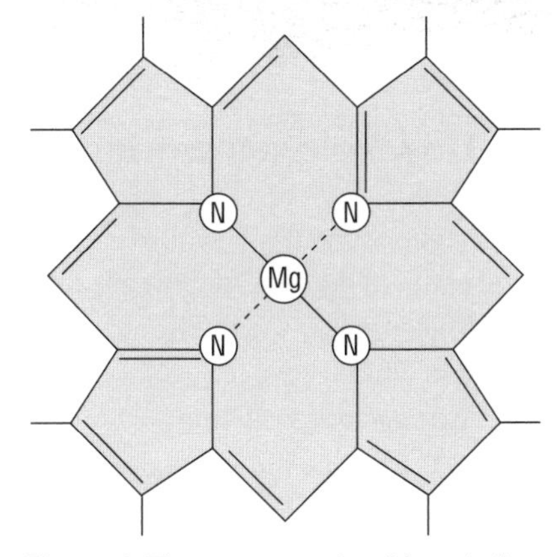

Figure 3 The structure of a chlorophyll molecule (simplified)

Examiner tip

You may need to compare the properties of radiation emitted by the Sun and by objects at room temperature (around 293 K).

Examiner tip: units of temperature and energy

Most questions about energy and radiant heat will give you temperature using the Kelvin scale. 0 °C = 273 K but otherwise the size of a 'unit' (1 °C or 1 K) is identical. So 25 °C = 298 K.

Energy is measured in joules (J), but you will often have to do calculations using the rate of transfer of energy, in J s^{-1} or watts (W).

Questions

1 Monastral blue is a pigment often used in blue paints. White light can be thought of as a mixture of blue, green and red light. What happens to **(a)** the blue light and **(b)** the red and green light in white light when it shines on the monastral blue pigment?

2 Explain, in terms of absorption and reflection, why freshly laid tar appears to be jet black.

3 How many times greater is the frequency of radiation given off by the Sun compared to that given off by the Earth?

4 Some stars are much hotter than the Sun – for example the bright star Vega, which can be easily seen from the UK overhead in summer, has a surface temperature of 9500 K. Suggest how the wavelength and frequency of the radiation given off by this star will differ from that given off by the Sun. What colour might Vega appear when we look at it in the night sky?

1.1 (5) Information transfer and the hazards of electromagnetic radiation

How science works

In this spread you will evaluate the possible dangers for human health caused by the different ways in which types of electromagnetic radiation interact with the human body (HSW 6a)

Scattering

A process which causes rays of light to change directions.

We are most familiar with visible light when it provides a way of transferring information – for example information from a piece of paper to our brain. This transfer of information involves several processes you have studied. The use of a camera is described here, but the processes would be very similar in the formation of an image in a human eye.

The formation of an image

As light falls onto a picture, some of it is absorbed by the dark pigments in the image and the rest undergoes **scattering** by the white sections of the image. If the image was on a smooth surface, such as a mirror, then simple reflection would occur rather than scattering. Entering the camera, the light is refracted, causing it to be focused on the back of the camera, where it forms an image. Here the light is absorbed by sensors which transfer the energy into electrical energy. In old, film-based cameras the light energy was transferred into chemical energy as silver compounds formed from the substances present in the film.

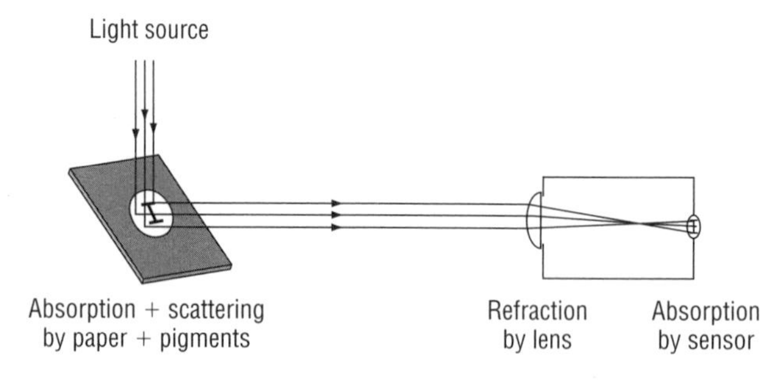

Figure 1 The formation of an image in a camera – the light rays from the original image are concentrated onto a very small region at the back of the camera

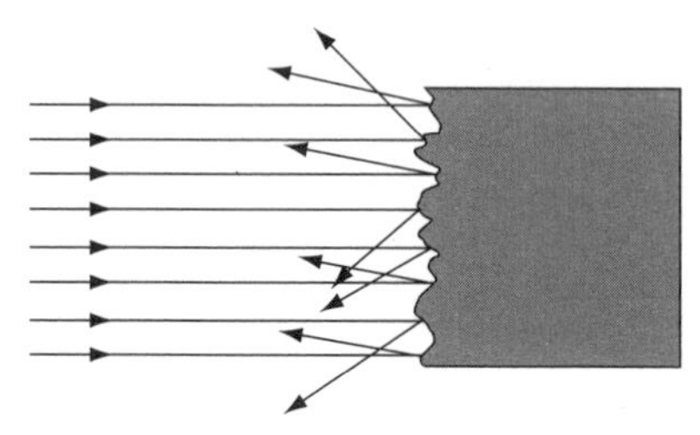

Figure 2 Illustration of scattering by a rough surface – unlike simple reflection, the rays of light are scattered in a range of directions

Limitations on the quality of the image

Refraction must focus the rays of light onto a very precise point at the back of the camera. To do this without blurring needs a very high-quality lens and a small aperture to limit the directions in which the light beams enter the camera. However, a small aperture can cause the problem of diffraction, as explained in spread 1.1.2 *The wave behaviour of light*.

Other types of information transfer

- Radio, TV and mobile phone communication, in which information about sound or images is added to the normal radio wave form, is either a digital or an analogue signal. This makes use of frequencies in the radio or microwave region of the spectrum.

AM radio	FM radio	Digital radio	TV	Mobile phone
150–1200 kHz	80–110 MHz	170–240 MHz	400–800 MHz	800–1900 MHz (extends into microwave region)

- The use of infrared radiation in remote control devices ('zappers') used to turn electronic equipment off and on. These transfer information – but also enough energy to activate a circuit in the receiver.
- Optical fibres are made of transparent glass or plastic and transmit light or ultraviolet radiation. As in radio broadcasting, the information is superimposed on the light or ultraviolet wave.

Digital

This refers to a signal in which information is transmitted in the form of a small number of possible numbers or levels – e.g. in the form of numbers from 0 to 255.

Analogue

A signal in which the information transmitted can take any value between certain limits.

- In barcode reading devices the pattern of visible light reflected by a bar code is converted into electrical signals which can be interpreted by a computer as information.

Electromagnetic radiation and human health

You have seen how some types of electromagnetic radiation interact with the human body:

- visible light – which is sensed by the eye
- infrared radiation – which causes a warming effect when it is absorbed by skin.

Other types of electromagnetic radiation also have effects on the human body, some of which may be damaging to health:

- Gamma rays and X-rays – these are at the high frequency end of the electromagnetic spectrum and carry the most intense energy. They are described as being 'ionising radiation' (see spread 2.4.5 *Radioactivity*) and so can cause damage to cells, including damage to DNA which may produce cancer.
- Ultraviolet radiation – this has a frequency slightly higher than visible light. Most ultraviolet radiation from the Sun is absorbed by the Earth's atmosphere (see spread 2.2.6 *Ozone and catalysts*) but when high frequency ultraviolet radiation (known as UVB) is absorbed by skin it can cause chemical bonds to break, causing sunburn and even skin cancer.
- Microwaves – although these have a much lower frequency than visible light, microwaves can cause rapid heating because molecules such as water absorb energy very effectively from microwaves.
- Radio waves – these are at the low frequency end of the spectrum and most scientists believe that they cannot cause harm to human beings. However, if emitters of radio waves are very powerful or held close to the body (as in the use of mobile phones) small heating effects in the body can be observed. Some people claim that this kind of exposure makes them feel ill and several large-scale studies have been carried out to assess the possible risk (see spread 2.4.12 *Epidemiology*).

Questions

1 Draw a diagram to show how diffraction of light waves can occur when they pass through a narrow aperture.

2 Explain why exposure to gamma radiation is considered much more damaging to human health than exposure to radio waves.

3 Calculate the wavelength of the highest frequency mobile phone signal – use the equation $c = f\lambda$ ($c = 3.0 \times 10^8\,\mathrm{m\,s^{-1}}$).

4 Sensors detect electromagnetic radiation and then produce an output signal which can be described as either digital or analogue. Classify these sensors as either digital or analogue:

   the human eye, a barcode reading device, the film used in photography

5 Processes which can happen to light include reflection, refraction, scattering, absorption and emission. White light is shone onto a black and white image on a piece of paper. A human eye looks at this piece of paper and sees an image. Describe the journey of the light to the back of the eye using some or all of these words to describe what happens to the light during the journey.

Examiner tip

You do not need to know details of how these processes work – you should simply to be able to give examples of how other types of radiation, apart from light, are used to transfer information.

Remote sensing

Satellites and environmental monitoring

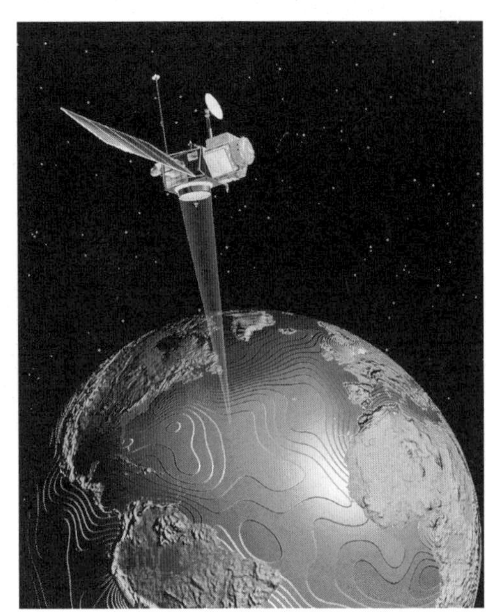

Figure 1 A satellite orbiting the Earth

Satellites have revolutionised scientists' ability to monitor the changes which are occurring to the Earth's surface, its oceans and its atmosphere. For example by **remote sensing** they can provide information about:

- weather patterns – enabling more accurate forecasting
- land use – for example the destruction of tropical rainforest can be monitored
- the surface temperature of the ocean – giving scientists early warning about, for example, changes in the pattern of ocean currents (see spread 2.1.6 *Ocean circulation*)
- the land surface of the Earth – allowing more accurate maps to be produced and a better understanding of its geology.

Using the electromagnetic spectrum

Satellite sensors use a wide range of **wavebands** from the electromagnetic spectrum to produce remote sensed images. These include:

- Visible light – often using several separate wavebands corresponding to a range of colours of visible light. Visible images show the visible light reflected from features on the Earth's surface or the atmosphere. The Sun is, of course, the original source of this visible light.
- Near infrared – a frequency range of infrared close to that of visible light. The Sun also emits near infrared and images made using this radiation show the amount of near infrared *reflected* from features. Healthy vegetation reflects near infrared strongly and water absorbs it strongly.
- Thermal infrared (radiant heat) – lower frequency than near infrared. This is *emitted* by warm or hot objects and so the thermal infrared detected by a satellite sensor will have been emitted (not reflected) by an object on the Earth's surface or in the atmosphere. The amount of thermal infrared emitted gives information about the temperature of objects (such as the tops of clouds).
- Radio waves – used in radar. These have a much lower frequency (longer wavelength) than the other types of radiation used. The key difference in the use of radar is that the satellite emits the radio wave and then measures the amount of radiation which is *scattered* from the surface (see spread 1.1.5 *Information transfer and the hazards of electromagnetic radiation*) at different angles. Radar can provide information about how rough a surface is, and also the moisture content of soil.

Remote sensing

Obtaining information about something without being in direct physical contact with it.

Activity

Search the Internet for images produced by remote sensing. NASA websites are often a very good place to start – try searching 'nasa + remote sensing' or 'nasa + satellite images'.

Find out about some of the information about the Earth which has been obtained from remote sensing satellites.

Waveband

Radiation with a range of wavelengths.

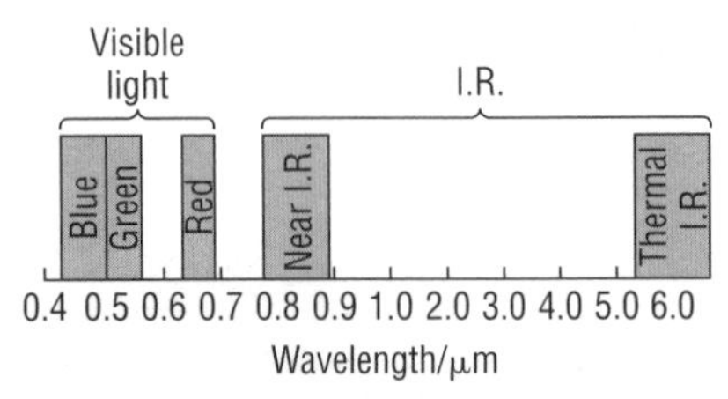

Figure 2 Wavebands used in remote sensing

a

b

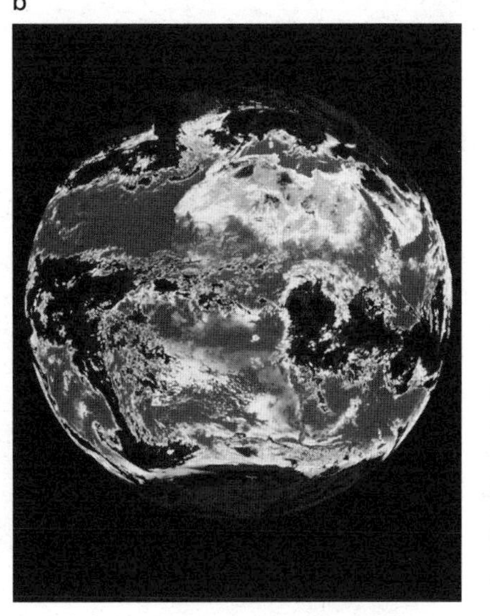

c

Figure 3 Remote-sensed images: **a** visible image from METEOSAT showing cloud formations; **b** positive thermal infrared image showing surface temperatures; **c** radar image of mountainous terrain

Displaying images

Grey-scale

Satellite sensors detect the amount of radiation emitted or reflected from a small section of the Earth's surface. The sensor converts this signal into a number (between 0 and 255) depending on the intensity of the signal. This number then determines the brightness of the pixel in the image.

240	28	73
240	240	125
125	125	125

Figure 4 An array of pixels showing how the numbers are converted into a grey-scale image

Pseudocolour and false colour

Grey-scale images are often hard to interpret because our eyes can detect only a small number of different shades of grey. In many cases a pseudocolour image is produced by allocating colours to different shades of grey. You will normally need to look at a key to see how the grey shades have been represented. Obviously these images are hard to show in a black-and-white book but there are many suitable examples on the Internet.

False colour images are rather morej16

complex to interpret. They provide a way of representing information about three different wavebands, detected by the sensor, in a single image. If the frequency ranges being detected are those of red, green and blue light, the image can be displayed in true colour and would look very much like what any astronauts aboard the satellite would see with their eyes. However, it is usually more important to display images which contain information about, for example, the amount of near infrared being detected. This means that the colours displayed in the image cannot correspond to those detected by the sensor (because one of them – infrared – is not visible to the human eye).

Questions on this part of the course can be found in the next two spreads.

Pixel

('*pic*ture *el*ement'). The smallest component of a graphic image. In remote-sensed images these are usually squares, representing information from a fixed area of land, e.g. 10 × 10 m.

False colour image

An image in which the colours displayed are different to the type of radiation being detected by the sensor.

The term may also be used to include *pseudocolour* (where colour is added to a black and white image) or *negative images* (where black and white are reversed).

Applications: false colour

Land-use images often use near infrared radiation to show the presence of healthy vegetation.

In false colour land-use images, information about radiation in the green, red and near infrared part of the spectrum is displayed. The colours used to do this are:

- green light (displayed as blue in the image)
- red light (displayed as green in the image)
- near infrared (displayed as red in the image).

Analysing remote sensed images (I)

Seeing beyond the visible: satellite thermal images

The images in Figure 1 come from a METEOSAT weather satellite. Its orbit is designed so that it remains above one point on the Earth's surface and can be used to monitor the weather patterns of a complete hemisphere. Other similar images can be obtained from suitable websites. The following activities will help you to analyse them. You may need to refer back to previous spreads in this module.

a

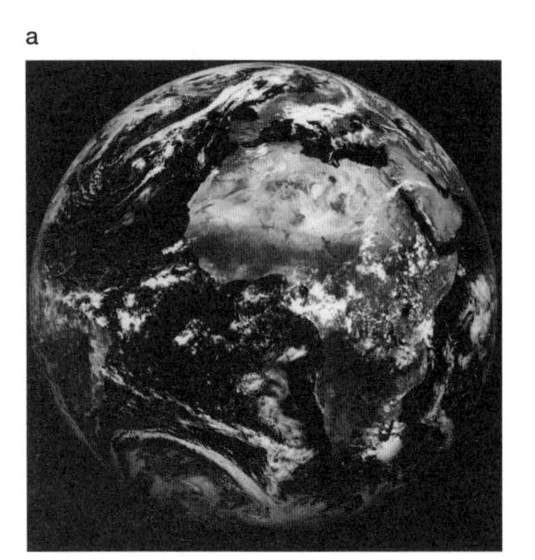

b

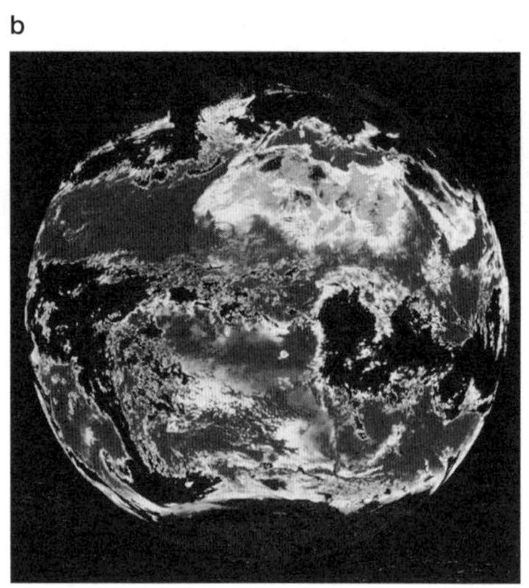

c

Figure 1 METEOSAT images: **a** in visible light; **b** using radiant heat; **c** using radiant heat but displayed in reverse

How science works

In this spread you will interpret grey-scale images of the Earth made using electromagnetic radiation (HSW 5b)

METEOSAT senses radiation reaching it in three wavebands – the wavelengths of the three wavebands are:

Band 1: 0.4–1.1 μm   Band 2: 5.7–7.1 μm   Band 3: 10.5 12.5 μm

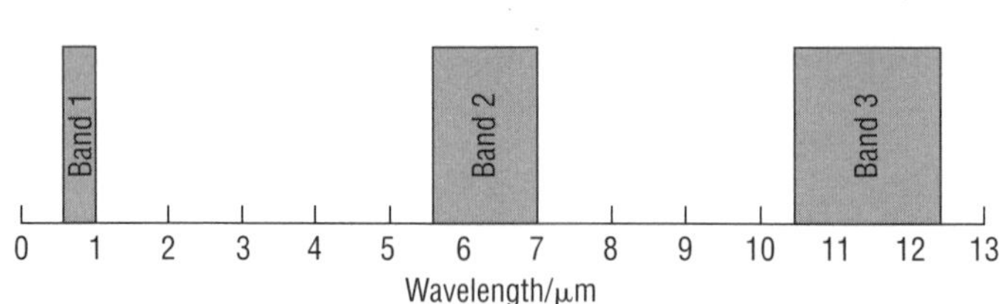

Figure 2 Representation of the three wavebands used by METEOSAT

1. Convert the wavelength ranges in Band 1 and Band 3 into standard form (See spread 1.1.1 *Light and electromagnetic radiation*).
2. Which parts of the electromagnetic spectrum contain Band 1 and Band 3? (See spread 1.1.1.) Suggest which band has been used to produce each of the images in Figure 1.
3. The radiant heat image is normally published in reverse (or negative) form, as in Figure 1c. Suggest why this is thought to be helpful.

However, when first analysing radiant heat images, you may find it easier to use the 'normal', unreversed form.

4 Use Figures 1a and b to compare the reflectivity (ability to reflect light) and relative temperatures of the features shown in the images. Copy and complete Table 1 in your notes.

Image A			Image B		
Feature	**Appearance**	**Reflectivity**	**Feature**	**Appearance**	**Relative temperature**
Cloud	White	High	Cloud		
Land			Land	White or pale grey	High
Sea			Sea		

Table 1

Clouds consist of tiny droplets of water and ice crystals. They can extend up to 10 km high in the atmosphere.

5 In what state would you expect water to be at the top of the cloud?

6 Use the information in Table 1 and your knowledge of clouds to explain the following points:
(a) clouds appear black or very dark grey in Figure 1b (infrared)
(b) clouds appear white in Figure 1a (visible light)
(c) some clouds in Figure 1b (e.g. those over the equator in Africa) appear darker than others.

7 Use the information in Table 1 and your knowledge of geography (alternatively find suitable information to help you) to suggest explanations for the following:
(a) the land in north Africa appears much lighter than the land in central Africa (Figure 1a)
(b) the land in Africa appears slightly lighter than the land in Europe (Figure 1b).

8 How would the appearance of Figure 1a and Figure 1b be different if they were obtained at night?

Examiner tip

You may be asked to interpret radiant heat images shown in negative (reverse) form. This adds an extra level of difficulty because you will need to remember that a dark area in a negative image would be light in the original image, and suggests that the area is emitting a lot of thermal infrared. Therefore it is a relatively hot region.

Try writing an explanation for why clouds appear white in a negative infrared image.

Satellite radar images

Radar images are helpful to view alongside visible or infrared regions because they show the relief (pattern of high and low ground) very effectively. In some cases radar provides the only way of obtaining detailed data about the relief.

Titan is one of the largest moons orbiting Saturn. In 2006 the space probe Cassini obtained data about its surface.

9 Figure 3 is a radar image of the surface of Titan obtained by Cassini. Find out why it is impossible to obtain visible images of the surface of Titan.

10 Suggest what the radar image tells us about the surface of Titan.

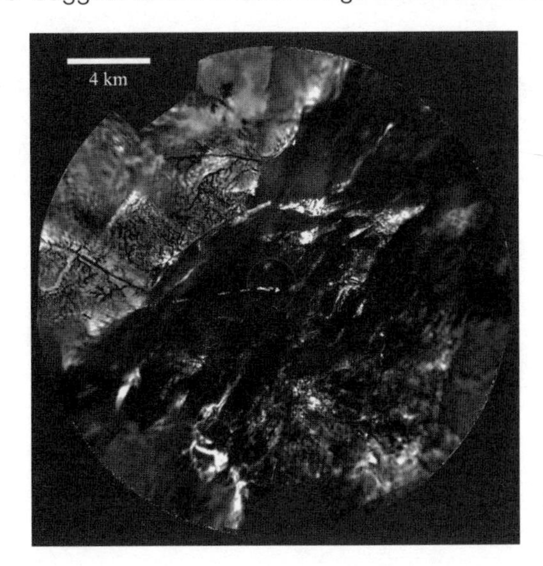

Figure 3 Radar image of a region of the surface of Titan

1.1 (8) Analysing remote sensed images (II)

Seeing beyond the visible

Figure 1 shows simplified diagrams based on data from a SPOT satellite. It orbits Earth at a much closer distance than the METEOSAT in spread 1.1.7. As a result, it can produce high-resolution images which are helpful in collecting data about land use, map making, etc.

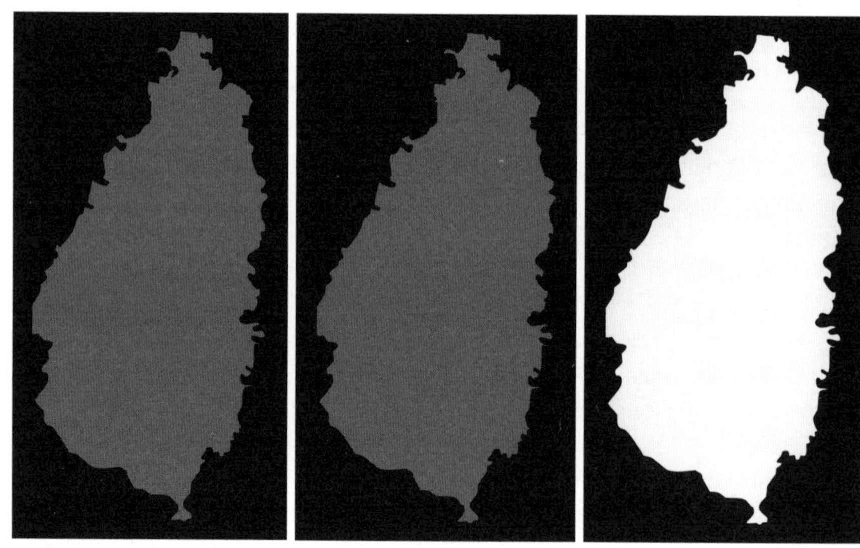

Figure 1 St. Lucia as it would appear using: **a** green light; **b** red light; **c** near infrared radiation

SPOT senses radiation reaching it in three bands – the wavelengths of the three bands are:

Band 1: 0.50–0.59 μm Band 2: 0.61–0.68 μm Band 3: 0.79–0.89 μm

1 Work out what type of radiation (green light, red light or near infrared) corresponds to each of these bands (see spread 1.1.1 *Light and electromagnetic radiation*). Note: Blue light is not very useful in studying Earth's surface because it is quite strongly scattered by the molecules in Earth's atmosphere.

2 St. Lucia is almost entirely covered in lush tropical vegetation, which is why the remotely sensed images have a very simple appearance. The top of the rainforest consists of a dense canopy of leaves. Copy and complete Table 1 to compare the reflectivity of the rainforest in the three wavebands.

	Appearance	Reflectivity
Green light	Dark grey	Low
Red light		
Near infrared		

Table 1

This explains why the near infrared image in Figure 1c appears so white. Images using this waveband are very helpful in monitoring the growth of vegetation.The use of false colour

Using the Internet

Many of the remotely sensed images available on the Internet are displayed in false colour and are difficult to reproduce in print. A good source of images is the Remote Sensing Tutorial, available at http://rst.gsfc.nasa.gov, or using a search engine to search for 'remote sensing tutorial'. Many greyscale and false colour images are displayed on the site, along with detailed information about remote sensing satellites and activities, allowing you to develop advanced skills in analysing the images.

The use of false colour

It is possible to display information from all three wavebands together in a false colour image. For example, the wavebands could be displayed in this way:

- a blue colour in the image is used to display information about band 1 (green light) received by the sensor
- a green colour in the image is used to display information about band 2 (red light) received by the sensor
- a red colour in the image is used to display information about band 3 (near infrared) received by the sensor.

The image on the cover of this book is an example of such a false colour image. It shows a region of the eastern Himalayas, northeast of Bhutan, China. The landscape consists of soaring mountains with snow-capped peaks. Vegetation grows on the lower slopes. Major rivers run through the valleys, fed by glacial meltwater. The area shown in the image is around 60 kilometres wide.

If you know the way in which the wavebands are displayed, it is possible to explain the appearance of the various features in the image.

You will need to remember several facts about the way that electromagnetic radiation is reflected:

- ice and snow reflect visible light and infrared strongly
- water itself does not reflect visible light or infrared very well
- mud or sediment will reflect certain colours of visible light well
- vegetation is normally quite dark so does not reflect visible light well. However, it reflects near infrared very well
- rock is often dark in colour so it will not reflect visible light or near infrared very well.

3 Copy and complete Table 2 to predict the appearance of some of the features which may appear on a false colour image of a Himalayan region, such as the one on the cover of this book.

Feature	Reflection of band 1 (green light) Displayed as *blue*	Reflection of band 2 (blue light) Displayed as *green*	Reflection of band 3 (near infrared) Displayed as *red*	Appearance in image
Sediment	High	Low	Low	Blue
Clear water	Low			
Snow			High	Blue + green + red = white
Vegetation		Low		
Bare rock				

Table 2

4 Use your predictions to interpret the image. Identify areas of:
  (a) snow
  (b) vegetation
  (c) bare rock.

5 There are two different types of blue area visible on the image – light blue and dark blue. Suggest what these features could be.

Examiner tip

To predict false colours in these composite images you will need to know the colours which are produced by mixing primary colours:

- blue + green + red = white
- blue + green = cyan (turquoise)
- blue + red = magenta (purple)
- green + red = yellow.

Remember that exam papers are normally only printed in black and white so you will most frequently be asked to look at a black and white map of a land area and predict the likely appearance of features.

Limitations of images

You have seen already how diffraction effects can limit the quality and sharpness of images obtained by the human eye or by cameras. There are several other factors which can limit the quality of images obtained by the sensors on satellites – or indeed any other detector.

The effect of the atmosphere

Satellites view the Earth's surface through the atmosphere – a layer of gases many kilometres thick. This causes three types of problem: reflection, scattering and absorption.

- Reflection – clouds prevent light and infrared radiation reaching the satellite because they are reflected from the lower surface of the cloud. Radar can normally penetrate clouds, although it may be scattered by large raindrops.
- Scattering – blue light is scattered by the molecules of nitrogen and oxygen in the air. Blue light is particularly affected because of its high frequency and short wavelength. As well as preventing reflected light reaching the satellite's sensors, this scattering process also makes the sky appear blue because the blue component of incoming sunlight is scattered most. Larger dust and water droplet particles are able to scatter green and red light as well, causing problems for satellite sensors but also giving the sky a milky-white appearance.
- Absorption – infrared and ultraviolet radiation are both absorbed by particular molecules in the Earth's atmosphere – ultraviolet by ozone (O_3) and infrared by greenhouse gases such as carbon dioxide and water vapour. The pattern of radiation reaching a satellite sensor is very different to the pattern of radiation emitted by (or reflected by) the Earth.

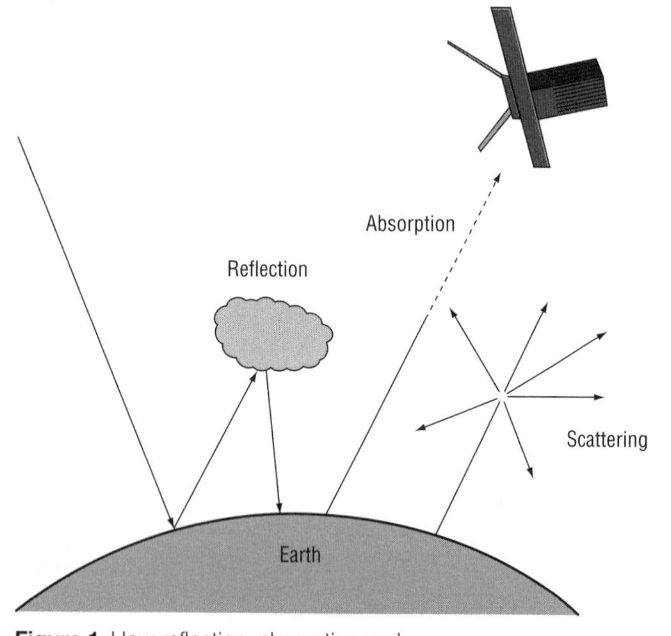

Figure 1 How reflection, absorption and scattering affect the radiation reaching a satellite-borne sensor

Figure 2 The red colours of sunset are caused by particles in the atmosphere scattering the blue and green light from the Sun

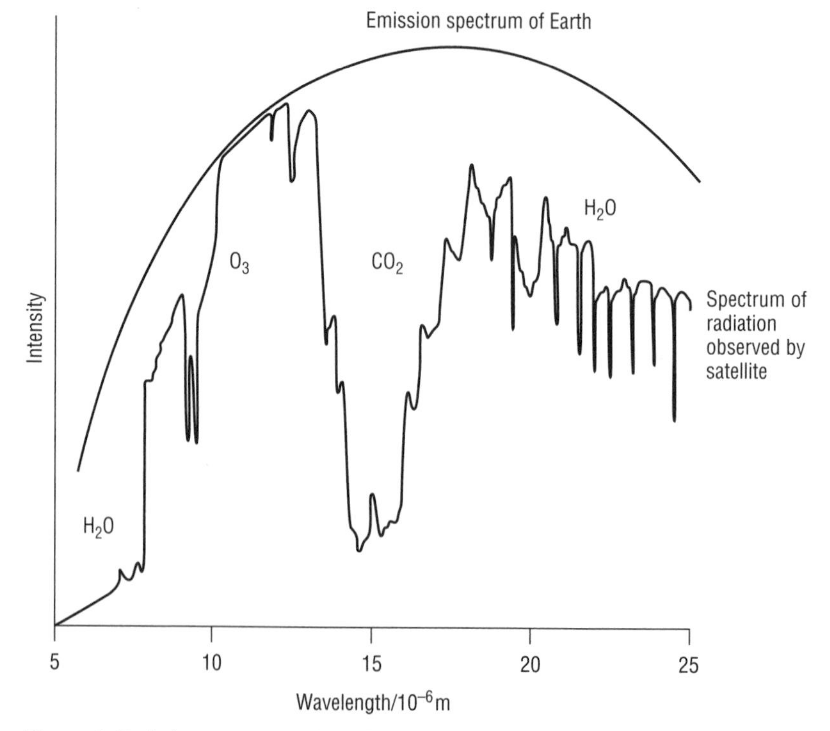

Figure 3 Emission spectrum of the Earth, compared with observed spectrum. The chemical formulas show the approximate wavelength of radiation absorbed by the gases in the atmosphere

Limitations of the sensor: pixels and resolution

Satellite sensors are in the form of an array of detectors, each of which records information about a rectangular (or square) area of ground. The **resolution** of the image depends on the area of the ground which is represented by a single pixel in the image. Small objects will be too blurred to identify.

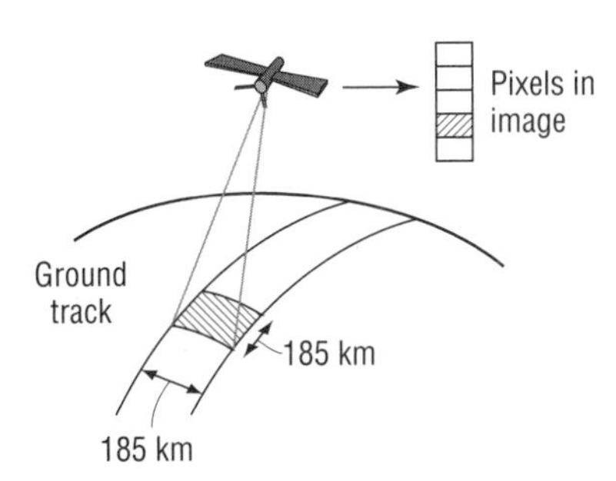

Figure 4 The relationship between ground area and picture element (pixel)

Limitations of the sensor: sensitivity and frequency ranges

The human eye

You have already looked in some detail at the frequency ranges to which cone cells are sensitive in the human eye.

There are also *rod* cells present in the eye. These are about 1000 times more sensitive than cone cells, so they allow the brain to produce an image at much lower light levels, such as in moonlight. There is only one type of rod cell, so colour vision is not possible. Although rod cells can detect frequencies across the whole range of visible light, they are more sensitive to green light – so it is much easier to see green objects at night. A grassy lawn will appear much lighter to your eyes in moonlight than red or blue objects.

Charge-coupled devices (CCDs) and photographic film

CCDs are the electronic components of sensors in digital cameras, bar code readers etc. They convert patterns of light into electrical signals.

An advantage of CCDs is that they can be designed to respond to any frequency range desired, unlike photographic film which is much more restricted in the way it can be used. Filters can, however, be used to remove unwanted frequencies of light, allowing images in, for example, 'yellow light' to be produced.

Questions

1 Blue light is rarely used in remote sensing of the Earth's surface. Why do the properties of blue light make it unsuitable for this kind of use?
2 The Sun appears red at sunset because only red light from the Sun reaches our eyes. What has happened to the blue and green light emitted by the Sun?
3 Why do some satellites (e.g. weather satellites) have a much lower resolution than others (e.g. land-use satellites)?
4 In spread 1.1.3 (*Light and colour*) you saw some graphs showing how the response of cone cells varied with frequency. Sketch a similar graph to show how the response of rod cells varies with the frequency of light.
5 Some land-use satellites have a resolution of about 2 m – each pixel in an image produced by the satellite represents about 4 m^2 of land.
   (a) If an image from the satellite contains 16 megapixels (1.6×10^7 pixels), calculate the area of the land displayed in the image.
   (b) Assuming that this is a square area of land, calculate the distance (in metres) from one side of the land area to the other.

Resolution

The smallest object which can be resolved (observed) on the ground.

Activity

Satellite mapping packages, such as Google Earth, show images of the whole of the Earth's surface but the resolution can be very different for different parts of the world. Use a suitable mapping package to look at your local area and estimate the resolution of the image. Hence, suggest what ground area is represented by one pixel.

Sensitivity

This is a measure of the intensity of an input signal (e.g. of a beam of light) required to produce some kind of output.

Activity: the vision of different organisms

1 Remind yourself of the different frequency ranges detected by some insects (such as bees) and by snakes (see spread 1.1.3 *Light and colour*). Suggest why these differences have evolved.
2 Find out about different types of eye – for example the eye of a fly compared to a human eye.
3 Birds of prey such as kestrels have remarkably sharp eyesight. Find out why.

1.1 Module 1 – Summary

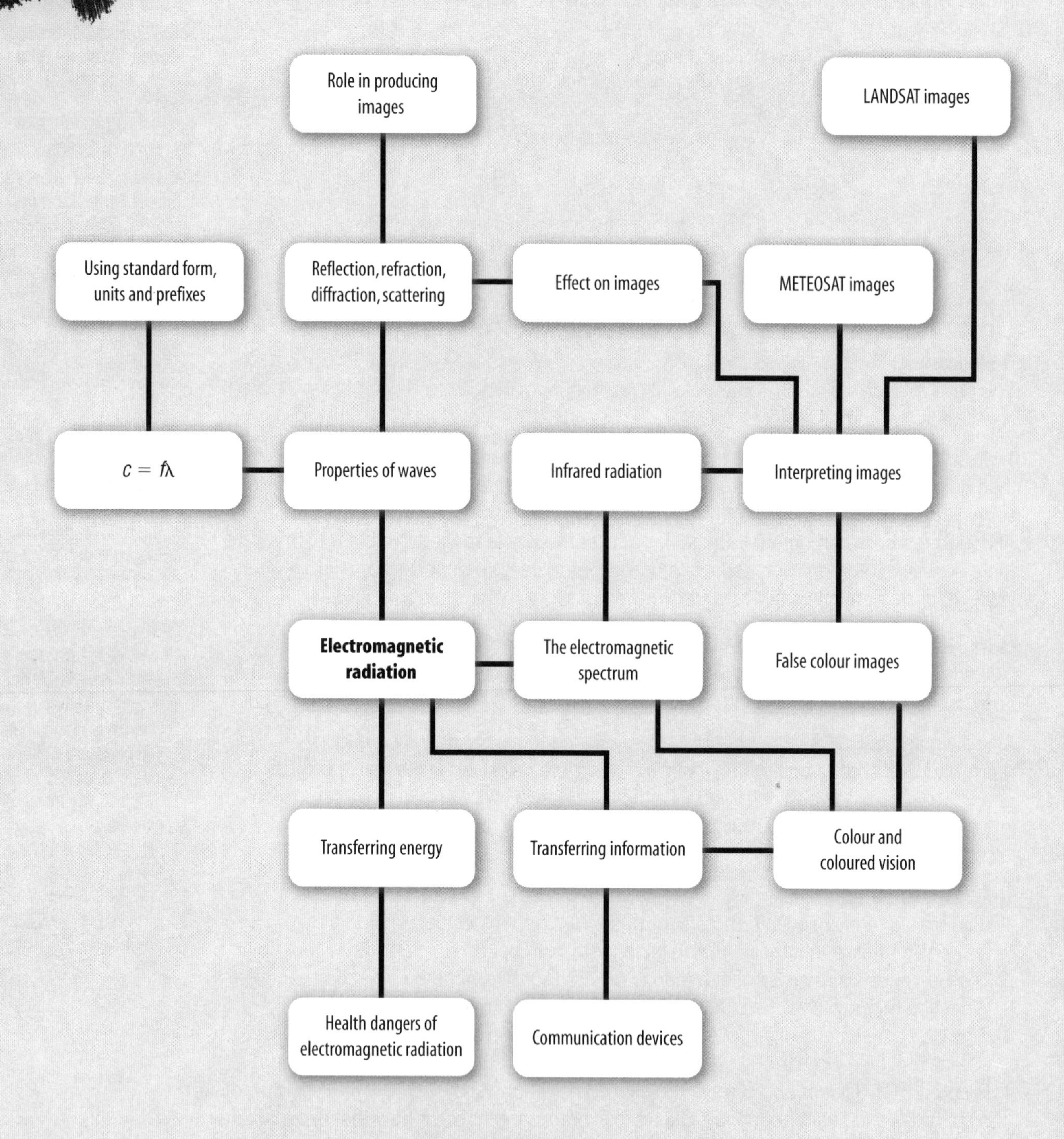

Practice questions

Low demand questions

These are the sort of questions that test your knowledge and understanding at E and E/U level.

1 The electromagnetic spectrum consists of a range of different types of radiation, differing from each other in their frequency and wavelength.

(a) (i) Place these types of radiation in order of their frequency, starting with the lowest:

ultraviolet radiation, radio waves, gamma radiation, visible light, infrared radiation

(ii) Which type of radiation is most likely to cause damage to human health? Explain your answer.

(b) Figure 1 shows three processes which all electromagnetic radiation undergoes. Name the three processes shown.

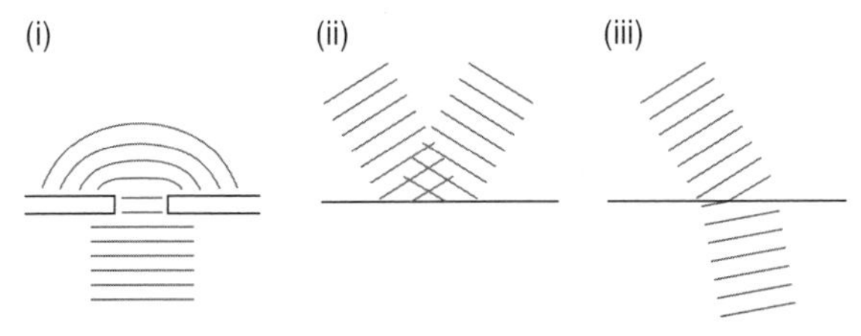

Figure 1a Three processes of electromagnetic radiation

2 Remote sensing satellites allow scientists to monitor changes in the environment on Earth (and other planets in the Solar System). They display information about the amount of radiation received in a particular waveband. This information is displayed in an array of pixels in an image.

(a) Explain the terms **(i)** waveband; **(ii)** pixel.

(b) METEOSAT satellites are used to monitor weather patterns on Earth. They use information about the radiation in several wavebands to produce images used by weather forecasters.

(i) Explain why clouds appear white in the visible image produced by METEOSAT.

(ii) Clouds appear black in the thermal infrared image produced by METEOSAT. Explain what you can deduce about the temperature of the tops of clouds from this information.

Medium demand questions

These are the sort of question that test your knowledge and understanding at C/D level.

3 Many satellites produce images that are described as 'false colour' images. In one such image, information in wavebands corresponding to green, red and near infrared radiation were displayed as a false colour image by assigning the following colours in the image to the three wavebands as shown in this table.

Waveband	Colour assigned in image
Green	Blue
Red	Green
Near infrared	Red

(a) Explain why this is described as a false colour image.

(b) The leaves of the trees in tropical rainforest reflect near infrared strongly but absorb most visible light. Ground which has been recently cleared of vegetation is pale in colour and reflects all three wavebands well. Predict the appearance in a false colour image of **(i)** tropical rainforest; **(ii)** recently cleared ground.

High demand questions

These are the sort of questions that test your knowledge and understanding at A/B level.

4 Digital radio signals are transmitted using frequencies of about 200 MHz.

(a) (i) Express 200 MHz in Hz, using standard form.

(ii) The equation which links frequency (f), velocity (c) and wavelength (λ) is $c = f\lambda$. Use this equation to calculate a value for the wavelength of digital radio signals.

(iii) Diffraction effects can sometimes occur when radio signals pass through gaps between high buildings, as shown in Figure 2. Use your answer to **(a) (ii)** to predict whether diffraction is likely to occur when digital radio signals pass through a gap approximately 100 m wide.

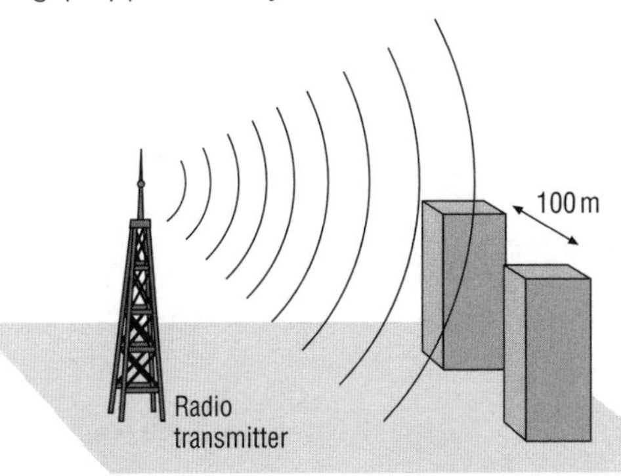

Figure 2 A radio signal passes through a 100 m gap

5 Coloured pigments, such as chlorophyll, are responsible for the green colour which we see when we observe most plants.

(a) Explain, in terms of the processes which occur when white light is shone on a green leaf, why the presence of chlorophyll makes leaves appear green.

(b) Describe what happens in the human eye which eventually results in our brain perceiving the colour green.

(c) What is different about the process which occurs if the light level is much lower, for example when plants are observed in moonlight?

UNIT 1

Module 2 Stable and vulnerable ecosystems

Introduction

In this module you will learn about the structure of ecosystems and how an understanding of ideas such as energy transfer and nutrient cycling can help us examine how sustainable and stable ecosystems are.

You will see how plants and other autotrophs have a crucial role to play in all ecosystems by trapping energy and transferring it into a useful form. The two key cellular reactions of photosynthesis and respiration will be studied in some detail as well as the mechanisms by which cells control the passage of substances into and out of the cell.

Having looked at the way in which ecosystems trap energy, you will remind yourself of the structure of ecosystems and how energy is transferred through the various feeding levels within it.

The rainforest is a particularly important ecosystem because of its biodiversity and its productivity. You will use ideas about natural selection and evolution to understand how this biodiversity came about and the potential impact of the destruction of the rainforest.

You will also see how nutrients, as well as energy, are passed through an ecosystem and how humans have used artificial sources of nutrients to increase productivity. The environmental impact of manufacturing and using these fertilisers is then examined.

Module contents

1 Autotrophs and photosynthesis
2 Energy transfer and active transport
3 Respiration and energy transfer
4 Ecosystems
5 Biodiversity
6 Natural selection and the development of new species
7 The nitrogen cycle
8 Nutrient fluxes and feedback
9 Fertiliser manufacture and eutrophication

How science works

During this module you will be covering some of the aspects of How Science Works. In particular you will be studying material which may be assessed for:

- HSW 5b: Analyse and interpret data to provide evidence
- HSW 6a: Consider applications and implications of science and appreciate their associated benefits and risks
- HSW 6b: Consider ethical issues in the treatment of humans, other organisms and the environment

Examples of this material include:

- Interpret data relating the productivity of a plant community to features of its environment (spread 1.2.4)
- Interpret data on the impact of human actions on biodiversity (spread 1.2.5)
- Describe industrial nitrogen fixing (spread 1.2.9)
- Describe the process of eutrophication as an example of disruption of nutrient cycling by human activity (spread 1.2.9)

Test yourself

1. What happens in the process of photosynthesis?
2. What happens in the process of respiration?
3. Give the names of two organelles found in all cells.
4. Explain what is meant by a food chain and give an example of a simple food chain, naming the organisms involved.
5. Name some nutrient elements – elements needed for healthy growth of plants.
6. Whereabouts in the world are tropical rainforests found?
7. What is the name of the process which causes new species of living things to develop?

Autotrophs and photosynthesis

Living organisms and cells

You should already know that:

- all living organisms are made up of *cells*
- complex organisms, such as plants or animals, consist of millions of cells
- there are a variety of structures within each cell, called *organelles*
- very simple organisms, such as bacteria, consist of single cells – these cells have a simple structure, so most of the organelles are not present
- all cells require energy to remain alive – they can transfer energy from a source into a usable form.

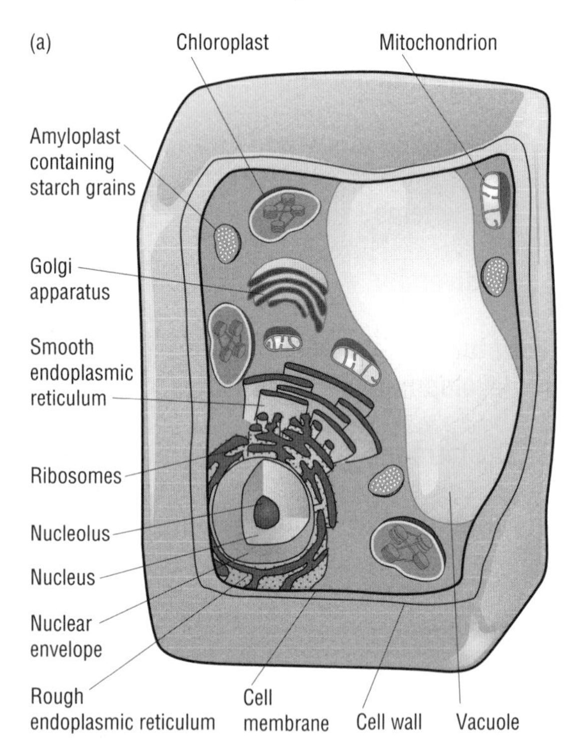

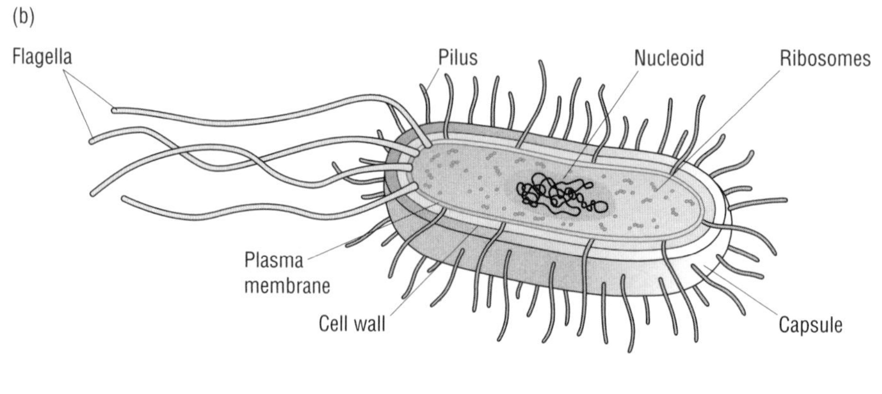

Figure 1a Plant cell – showing organelles; **b** bacterial cell

Examiner tip

You don't need to know the details of cell structure. However, you should be able to describe the role of the cell membrane, mitochondria, chloroplasts and cytoplasm in energy transfer (see spread 1.2.3 *Respiration and energy transfer*).

Biomass

The organic matter which makes up living organisms.

Biosynthesis

Reactions in living organisms which build up complex organic substances from simpler ones.

Autotroph

An organism that produces complex organic compounds from inorganic compounds and an external source of energy.

Examiner tip

You do not need to know the balanced chemical equation, but you should be able to see how it shows that the reaction is a biosynthesis reaction.

The reactions occurring in cells involve complex molecules described as *organic* molecules. These are based on long chains or rings of carbon atoms. Most organisms, and the cells within them, use organic molecules such as glucose as the source of their **biomass** – and also as the energy source to produce this biomass by the process of **biosynthesis**.

However, some organisms can use *inorganic* compounds (such as carbon dioxide and water) to produce biomass. They also use energy sources which are not derived from biomass. These organisms – called **autotrophs** – are essential to the existence of all ecosystems.

Photosynthesis

Photosynthesis is a biosynthetic reaction which occurs only in plants and some other organisms such as algae. It is the only way in which the *light energy* from sunlight can be used to produce biomass, which is a store of *chemical energy*.

The overall word equation is:

carbon dioxide + water → glucose + oxygen + energy

It can also be represented by a balanced chemical equation:

$$6CO_2 + 6H_2O \rightarrow C_6H_{12}O_6 + 6O_2 + \text{energy}$$

Light energy is trapped by a pigment known as chlorophyll. This absorbs mostly red and blue light – it reflects green light. You learned about this process in spread 1.1.4 *Absorption of light and energy transfer*.

There are two main stages in photosynthesis – although each stage is really made up of many quite complex steps:

- Light-dependent stage – light is absorbed by chlorophyll and is used to split molecules of water into oxygen molecules and hydrogen atoms. A molecule of ATP is also formed from ADP and inorganic phosphate ions (P_i).
- Light-independent stage – hydrogen atoms from the light-dependent stage are combined with carbon dioxide to form glucose. The energy for this process comes from the breakdown of the ATP molecule produced in the light-dependent stage.

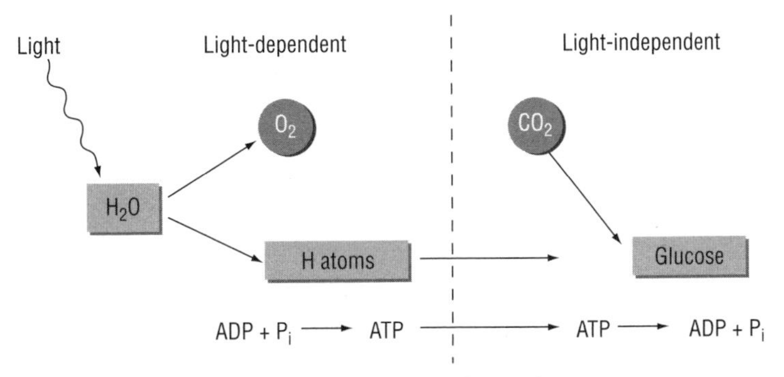

Figure 2 The two stages in a photosynthesis reaction

ATP

This is the abbreviation used for a molecule called adenosine triphosphate. Energy is used to form ATP from ADP (adenosine diphosphate) and a phosphate ion (P_i). Molecules of ATP can move around a cell easily from one organelle to another.

This energy can then be released instantly when ATP is broken down into ADP. So the formation and breaking down of ATP provides an easy way of transporting energy to where it is needed.

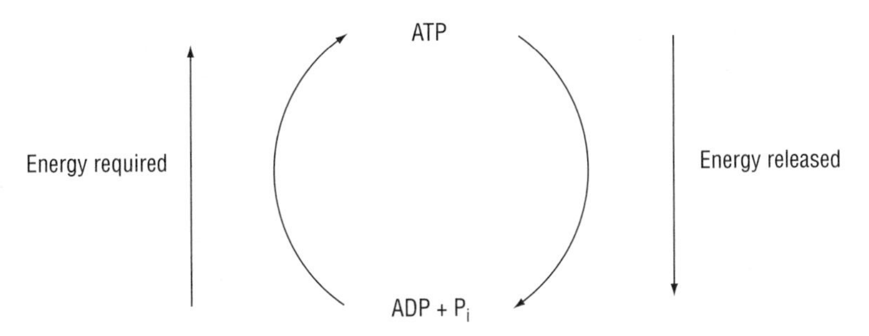

Figure 3 Formation and breakdown of ATP

Other autotrophs

Plants are the most obvious examples of autotrophs. However, some bacteria contain chlorophyll and they can also carry out the process of photosynthesis.

Also, bacteria have been discovered which do not use light as a source of energy – they seem to make use of the energy produced by reacting simple sulfur or iron compounds with oxygen. They are known as *chemautotrophs* because they obtain their energy from the chemical reactions of simple inorganic substances. However, the carbon which makes up the biomass of chemautotrophs must be obtained from other organic compounds.

Iron bacteria can cause problems in pipework made of iron or steel because they oxidise iron and iron compounds to iron oxide – this is similar to the substance we call rust. Sulfur bacteria are often found close to volcanic springs because volcanoes are a good source of sulfur compounds – one remarkable type of ecosystem that is dependent on these type of bacteria is based around massive underground thermal vents found deep under the ocean.

Activity

Chemautotrophic bacteria have been extensively studied by biologists because it is thought that they may well hold to key to how life began on Earth. It is also likely that if there is life elsewhere in the Solar System it would be chemautotrophic.

Try searching for 'chemautotrophic + Mars' or 'chemautotrophic + early life' to find out more. The ecosystems based on the bacteria around hydrothermal vents are also well worth exploring – try searching for 'hydrothermal communities'.

Questions

1 What is the difference between a plant and a chemautotroph?

2 Write word equations for these processes:
  (a) the overall photosynthesis reaction
  (b) the light-dependent stage of photosynthesis
  (c) the light-independent stage of photosynthesis.

3 Explain how ATP transports energy around a cell. Give an example of a reaction which requires energy from the breakdown of ATP.

4 'All cells require a source of energy'.
  (a) List three different sources of energy mentioned in this spread.
  (b) Give one reason why all cells require energy.

1.2 (2) Energy transfer and active transport

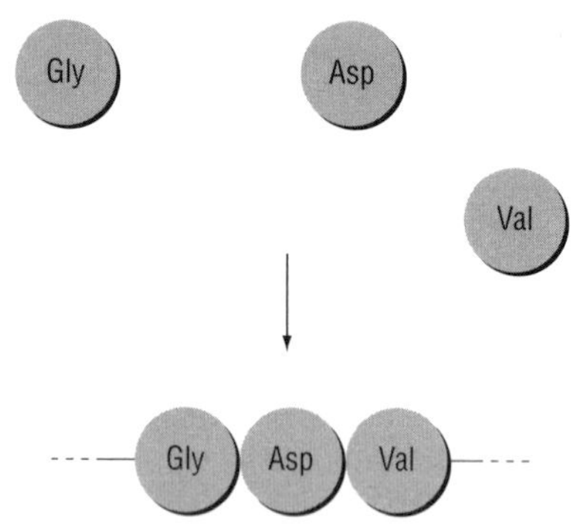

Figure 1 Biosynthesis of a protein

Why do cells need energy?

Autotrophs, such as plants (see spread 1.2.1 *Autotrophs and photosynthesis*), can transfer energy from light (or other non-living energy sources) and store it as chemical energy in cells. This chemical energy can then be used for a range of processes that are vital for the cell and the organism as a whole.

All other organisms obtain their energy from other living organisms. For example herbivores eat plants and use the energy stored in the biomass of the plant.

Organisms need a source of energy for several activities:

- biosynthesis – building up complex molecules from simpler ones, for example making proteins from amino acids
- movement – such as the contraction of muscles, or the beating of the hairs (cilia) of simple one-celled organisms
- moving molecules in and out of cells – a process known as active transport.

Passive and active transport

A huge range of molecules and ions must cross cell membranes in order to enter and leave cells. These include:

- small molecules – such as water, oxygen and carbon dioxide
- large molecules – such as glucose
- charged ions – such as sodium (Na^+), potassium (K^+) and chloride (Cl^-).

The cell membrane has a structure known as a fluid mosaic. Large molecules known as phospholipids are arranged in a double layer and protein molecules are embedded within this layer.

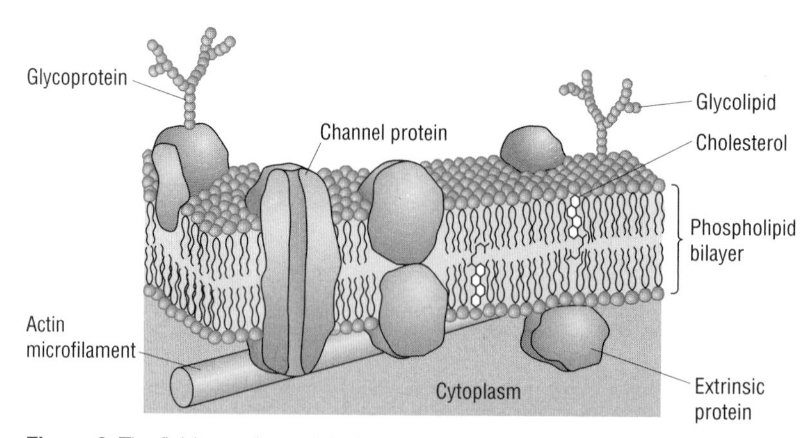

Figure 2 The fluid mosaic model of a cell membrane, showing the embedded protein channels

Passive transport

Small molecules – such as water, oxygen and carbon dioxide – can pass through the gaps between the lipids and move freely in and out of the cell. They will always move in a direction from high concentration to low concentration. They are said to be moving 'down a *concentration gradient*' – this is known as passive transport as no energy needs to be transferred in order for the process to occur. It is rather like rolling a heavy stone downhill.

Other molecules – such as glucose – are too large to pass through the bilayer but can pass through the protein channels in the membrane. The glucose also moves down a concentration gradient, and so this is also passive transport.

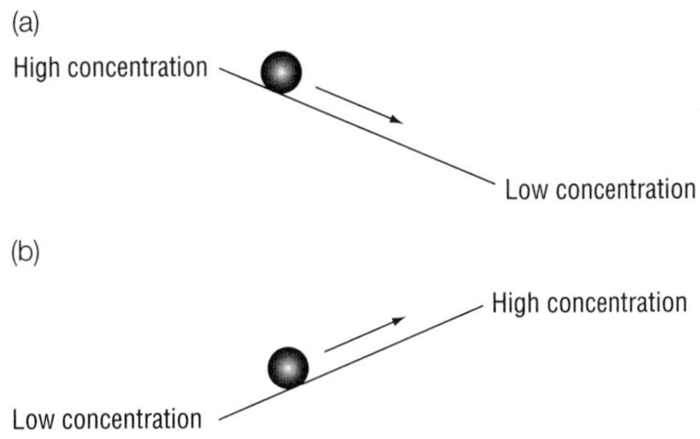

Figure 3a Moving down a concentration gradient does not require an input of energy; **b** moving up a concentration gradient requires an input of energy

Active transport

For some cell processes, it is necessary to move molecules or ions against a concentration gradient – this is like trying to push a heavy stone uphill. In this case, energy needs to be transferred to push molecules through the protein channels. The energy is provided by the breakdown of ATP (into ADP and P_i).

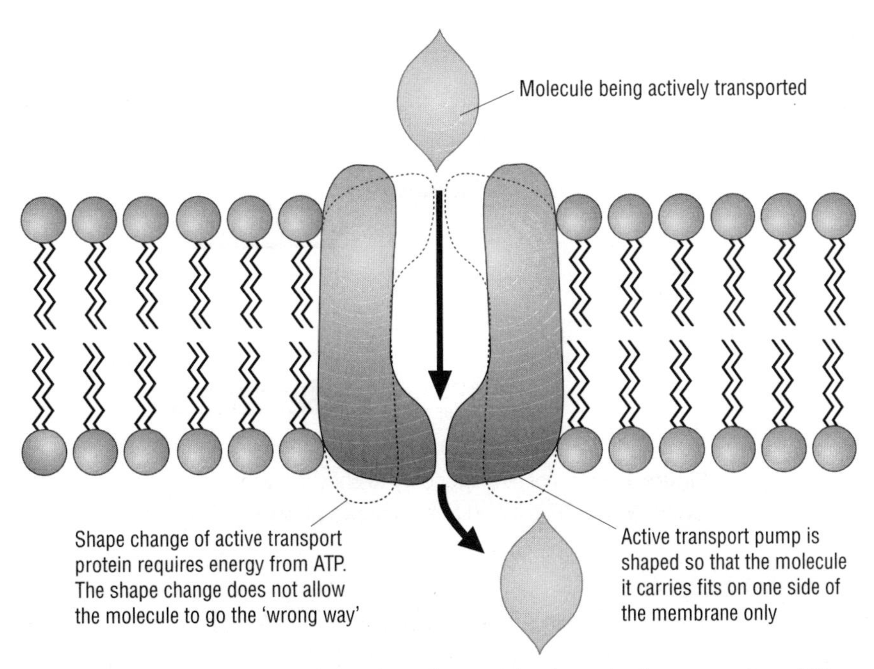

Figure 4 Movement of a substance through a protein channel using energy from ATP

Questions

1 Classify each of these processes as *biosynthesis*, *active transport* or *movement*:
  (a) pumping sodium ions out of a cell
  (b) converting small glucose molecules into large starch molecules
  (c) the beating of a heart.
2 Draw a diagram to show the structure of a cell membrane. Use your diagram to explain how **(a)** oxygen molecules, **(b)** glucose molecules and **(c)** chloride ions can pass through the membrane.

Respiration and energy transfer

Respiration

Autotrophs transfer light energy into chemical energy, which is stored in molecules such as glucose. In order to make use of this stored energy it must be transferred into a useable form by chemical reactions known as **respiration**.

Respiration can be *aerobic* (using oxygen) or *anaerobic* (not using oxygen). In aerobic respiration:

glucose + oxygen → carbon dioxide + water + ATP

$$C_6H_{12}O_6 + 6O_2 \rightarrow 6CO_2 + 6H_2O$$

The energy from the reaction also forms approximately 30 molecules of ATP.

In anaerobic reactions, other products, often toxic, are produced (such as ethanol or lactic acid) and many fewer ATP molecules are produced – typically only 2.

ATP

ATP is a mobile energy store (see spread 1.2.1). It can enter proteins, such as enzymes or the carrier proteins in the membrane, and transfer energy for use in biosynthesis or active transport. When it does this it breaks down to ADP (adenosine diphosphate) and a phosphate ion, P_i.

During respiration, ATP is formed from ADP and phosphate ions:

$$ADP + P_i \rightarrow ATP$$

This process requires energy – this is transferred from molecules such as glucose by the process of respiration.

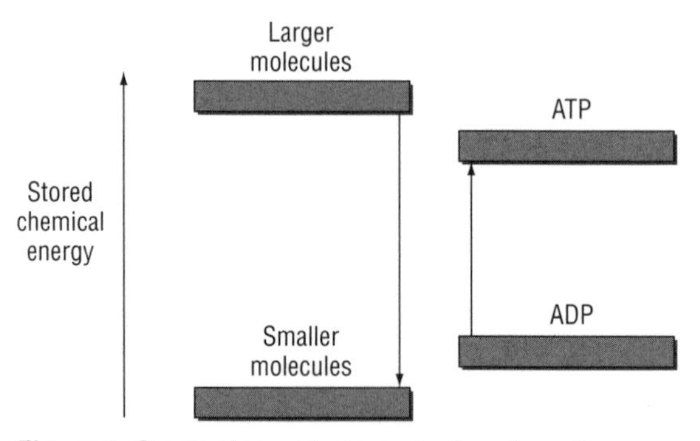

Figure 1 Graph of stored energy to show how the formation of ATP is linked to breakdown of glucose

The sites of respiration

Respiration takes place in two main stages. The glucose molecule is gradually broken down into carbon dioxide and some ATP molecules are formed in each stage.

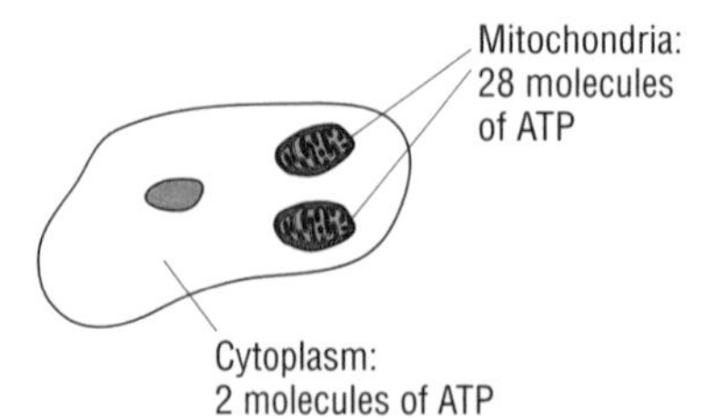

Figure 2 A cell showing mitochondria and cytoplasm

Respiration

A chemical process in which glucose is broken down into smaller molecules releasing energy (which is transferred to ATP molecules).

Applications: calculations to do with respiration

Aerobic respiration releases approximately 2800 kJ of energy from every mole* of glucose.

To form ATP from ADP requires approximately 30 kJ of energy for every mole of ATP.

Producing 30 moles of ATP requires 900 kJ of energy. The remaining 1900 kJ is dissipated as heat (see below).

*The mole is a fixed number of molecules of a substance and is a unit often used by chemists when comparing equal amounts of substances.

The function of organelles

Table 1 summarises the role of some of the organelles in the processes of energy transfer.

Organelle	Function	Role in energy transfer
Plasma membrane	Controls which substances enter and leave the cell	Allows glucose and oxygen into the cell; allows carbon dioxide out
Mitochondria	Site of respiration	Respiration transfers most of the energy stored in glucose into ATP (and some heat)
Chloroplasts (in plants and other autotrophs)	Site of photosynthesis	Chloroplasts trap the light energy from the sun and transfer it into chemical energy in the form of glucose
Cytoplasm	Site of some respiration reactions and most other chemical reactions	Respiration (see above); the ATP made in respiration diffuses through the cytoplasm to provide the energy required for a range of chemical reactions

Table 1 The role of organelles in the processes of energy transfer

Law of conservation of energy

Energy cannot be created or destroyed – only transferred from one form to another.

Energy transfer

All living systems must obey the **law of conservation of energy** – also known as the first law of thermodynamics.

Energy-flow diagrams can be used to represent the flow of energy in biological processes – for example between organisms in a food chain or within the cell when respiration is used to provide energy for cell processes.

There are two things to be aware of:

- the total amount of energy always remains unaltered
- in any process, some of the energy is converted into heat energy – this is known as the dissipation of energy.

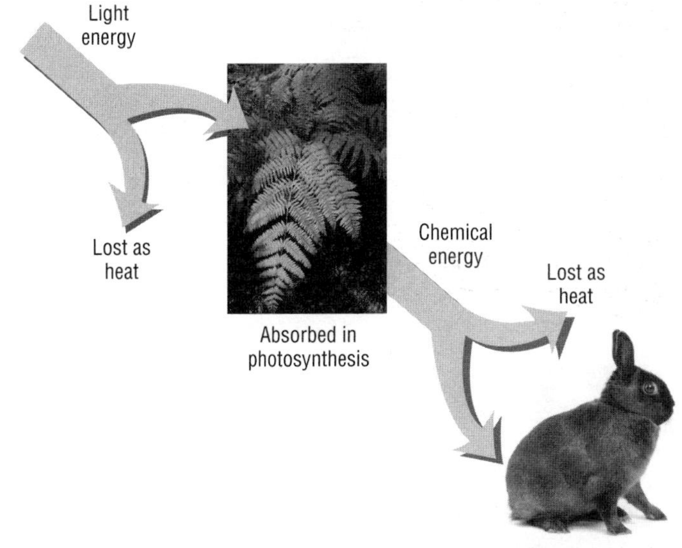

Figure 3 Energy flow for light energy being trapped by a plant cell and passed on to an animal

Questions

1 Name two organelles involved in the process of respiration.

2 Give two reasons why anerobic respiration is normally undesirable.

3 A total of 3.4×10^6 kJ of light energy reaches every 1 m^2 of a woodland ecosystem each year. 75% of this energy is reflected by the leaves or the ground. The rest is absorbed by the leaves.

(a) Calculate the amount of energy absorbed by the leaves.

(b) 42 500 kJ are transferred as stored chemical energy by photosynthesis, the rest is transferred as heat. Calculate the percentage of the absorbed light energy which is transferred as chemical energy.

(c) This stored energy is transferred in respiration to produce 400 moles of ATP. Each mole requires approximately 30 kJ of energy. Calculate the amount of energy transferred into the chemical energy of ATP.

(d) Calculate the amount of energy transferred as heat during respiration. What percentage of the stored chemical energy is dissipated as heat?

(e) Draw an energy-flow diagram to summarise the information given in the question.

Examiner tip

You may be asked to complete energy-flow diagrams or to calculate the amount of energy transferred in a particular process.

Ecosystems

In previous spreads you saw how energy is transferred from sunlight to autotrophs, and then looked at some of the ways that the energy might be used in organisms.

Scientists also study living things on a much larger scale, to understand how these organisms interact with each other – this branch of science is known as *ecology*.

Ecosystems can be as simple as a small pond (or even a fish tank) but can also represent vast areas of the Earth – for example a rainforest or a desert. These are often known as **biomes**.

Food chains and food webs

These are used to show the feeding relationships in ecosystems – in other words, how chemical energy is transferred between the different **populations** of organisms which make up the **community** occupying the **habitat**.

Autotrophs (usually plants) are commonly shown at the bottom of the food chain or web. These are called producers – the organisms above the plants in the chain or web are called consumers.

Not all the energy from biomass is passed on to other organisms in the food chain. Some is dissipated as heat, some is lost in excretion as faeces – or the animal may die and decompose without being eaten by other organisms. Biomass lost in excretion or decomposition is broken down and eventually transferred as heat by decomposer organisms such as bacteria and fungi.

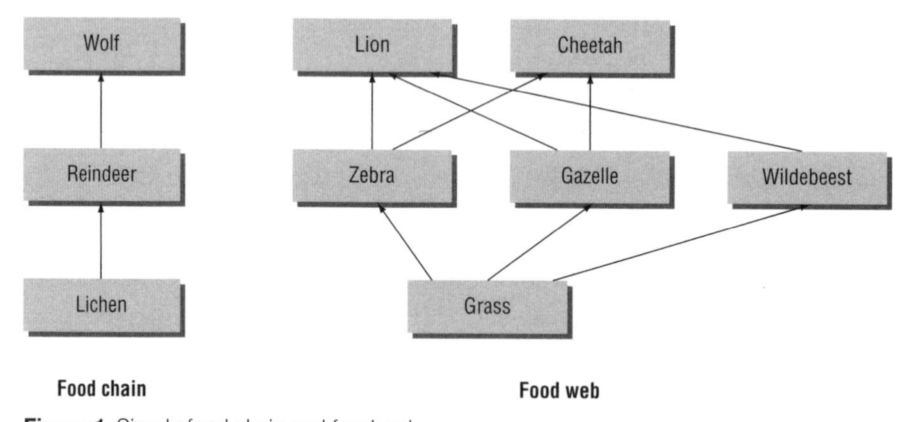

Figure 1 Simple food chain and food web

Examiner tip

You will not need to recall any details of the structures of food chains or food webs, but you may be asked to perform calculations connected with the energy transferred from one organism to other organisms in a food chain or web.

Productivity

Plants and other autotrophs are described as being the producers of an ecosystem because they are the only organisms which can trap energy from non-living sources and make it available to the other organisms in the ecosystem.

Productivity varies greatly between ecosystems – one reason for this is the climate of each ecosystem. Table 1 shows some typical values of productivity for some ecosystems along with some data about the climate.

Ecosystem

The community of living organisms interacting with each other and the physical environment in which they live.

Biome

A very large region consisting of similar ecosystems.

Activity

The Earth can be divided into six different biomes (some scientists suggest seven or eight). Find out about these biomes – for example the physical conditions of the habitats and the ways in which organisms have adapted to survive. Simply search the Internet for 'biomes'.

The BBC's *Planet Earth* series made in 2006 was divided into episodes, each dealing with a different type of ecosystem. Video clips from the series are available on the BBC website.

Population

The organisms of a particular species living in an ecosystem.

Community

The living organisms in a particular ecosystem.

Habitat

The physical environment in which a community lives.

Productivity

The amount of energy trapped by an ecosystem in the form of biomass (organic matter).

It is normally defined as the amount of energy (in kJ) trapped by 1 m^2 of land in a year. The units of productivity are, therefore, kJ m^{-2} yr^{-1}.

Ecosystem	Productivity/ kJ m^{-2} yr^{-1}	Average temperature/°C	Rainfall/mm	Solar radiation/ kJ cm^{-2}
Tropical forest	2200	27	1500	650
Temperate forest	1200	12	500	550
Boreal forest	800	0	375	400
Savannah	900	24	425	750
Temperate grassland	600	12	325	600
Tundra	140	−10	140	300
Desert	90	24	100	800

Table 1 Productivity in ecosystems in different climates

In reality, the overall productivity will be the result of an interaction between these and other factors – for example the type of soil will also make a big difference. Some soils are rich in nutrients and hold water well – both factors will help to increase productivity. The average temperature will depend not only on how close the ecosystem is to the equator but also on the altitude (height above sea level). Mountainous regions often have unique ecosystems because of unusual combinations of climatic factors.

The future of the productivity of ecosystems

The productivity of the ecosystems of the world may change in the future because of global warming, which will alter patterns of temperature and rainfall. In addition, the increased carbon dioxide level, which is the likely cause of global warming, may also affect productivity – probably increasing it. Some scientists believe that the overall effect of all these changes may be to increase the overall productivity of the Earth, which will then remove more carbon dioxide from the atmosphere. This could help to slow down global warming in a process known as *negative feedback*. Other scientists are not so sure and suspect that productivity will drop – then less carbon dioxide will be removed from the atmosphere and global warming will accelerate. This would be a *positive feedback* effect. For more about feedback effects see spread 1.2.8 *Nutrient fluxes and feedback*.

How science works

In this spread you will interpret data relating the productivity of ecosystems to a variety of climatic factors (HSW 5b)

Activity

To see which factors are most closely linked with productivity, try using a spreadsheet to produce separate scatterplots of *productivity* against *temperature, rainfall* and *solar radiation* and look for evidence of any patterns.

Questions

1 Table 2 summarises information about various words connected with ecosystems. However, the words do not match the descriptions and examples as written. Rewrite the table to correctly summarise what you know about these key terms.

Key term	Definition	Example
Ecosystem	A collection of different living organisms	The sand in the Sahara desert and the harsh climate which created it
Biome	Community of organisms + the habitat	The Canadian arctic tundra
Community	All the organisms of one species	The plants, fish and insects living in a small fish tank
Population	The physical environment	A small lake in Scotland
Habitat	Many similar ecosystems covering a large area	Emperor penguins living on the island of South Georgia

Table 2

2 Suggest reasons for:
(a) the low productivity of the tundra biome
(b) the high productivity of the tropical rainforest biome

3 Which ecosystem in Table 1 will have the simplest food chains? Explain why.

1.2 (5) Biodiversity

How science works

In this spread you will consider the implications for biodiversity of the deforestation of rainforest ecosystems (HSW 6b).

Tropical rainforest – a uniquely valuable habitat

One of the biggest environmental concerns in recent decades has been the rapid destruction of tropical rainforest. Since 1950 the area of the Earth's tropical rainforest has fallen by about 50%, and in some parts of the world, such as south-east Asia, by nearly 70%.

Tropical rainforests have an exceptionally high productivity – they absorb a lot of carbon dioxide from the atmosphere. Once cleared, the rainforest is replaced by grazing land with a much lower productivity. Less carbon dioxide is absorbed and the grazing animals release methane – another potent *greenhouse gas*. In order to clear the forest, the trees are simply burned releasing all the carbon in their biomass as carbon dioxide. So the overall effect is to accelerate the increase in the carbon dioxide concentration in the atmosphere, causing increased global warming. You will learn more about greenhouse gases and global warming in many of the other spreads in this book.

But there are more environmental consequences than that. The rainforest has the highest **biodiversity** of any ecosystem.

There are estimated to be 5 million different species of plants, animals and insects living in the tropical rainforest – more than half of the world's total. One hectare may contain over 750 types of trees and 1500 species of higher plants.

Not only is much of the habitat being lost, but the fragmentation of the rainforest has made the remaining habitat too small in area to support many of the larger animals at the top of the food chain. The ecology of a rainforest is particularly intricate and fragile, and small changes to one part of the system can create unexpected and devastating effects on the whole ecosystem.

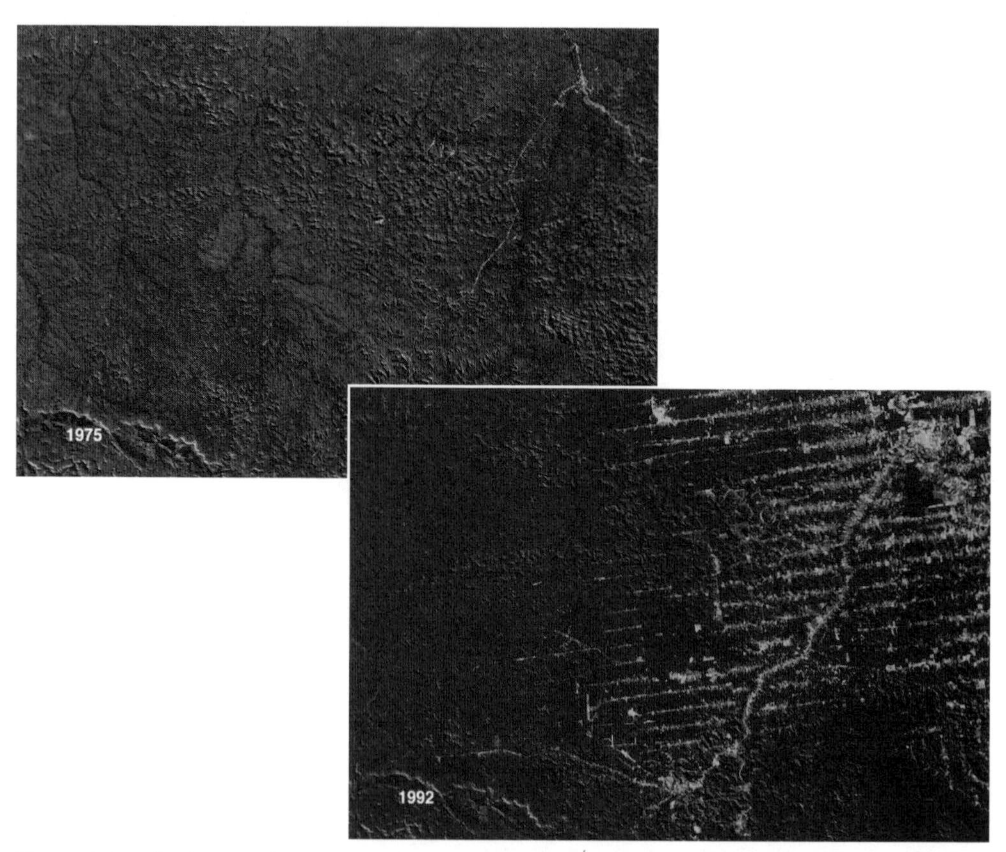

Figure 1 Satellite images showing destruction of rainforest between 1975 and 1992. The pale grey areas in the 1992 image show where forest has been cleared either side of long straight roads.

Biodiversity

The variety of organisms present in an ecosystem. This is often measured as the number of different **species** present, but it can also take into account the different varieties of organisms within a species.

Species

A separate group of organisms that can successfully interbreed with one another.

(Note: there is often disagreement between scientists about whether organisms are different species or simply varieties of the same species.)

Activity

There are numerous sources of satellite images on the Internet which can illustrate the speed at which rainforest destruction is happening. Search for 'rainforest + satellite images'. Google Earth will also give you an idea of the dimensions of, for example, the Amazon rainforest.

Several of these sites also show the extent of some of the fires which have been used to burn the rainforest prior to clearing the ground.

Why is the rainforest so high in biodiversity?

There are a variety of reasons which have been identified:

- The rainforests are very old ecosystems which may have occupied the same geographical areas for millions of years.
- The rainforests have very high productivity, allowing more extensive food chains and food webs.
- The rainforests have a complex structure with a great variety of locations – or *niches* – within the ecosystem that can be occupied by different organisms. For example completely different food webs exist in each layer of the rainforest.

Why does loss of biodiversity matter to us?

Apart from the fact that the rainforest ecosystem may be affected in unpredictable ways by the loss of biodiversity, many of the species in the rainforest – particularly plant species – have important uses to human society.

Plant species in the rainforests have been a particularly rich source of pharmaceutical products. Only a tiny fraction of the plant species has been investigated in this way up until now. Extinction of these species could potentially remove potent medicines for the future forever.

Many of the most popular components of the developed world's diet originated in the rainforest, including bananas, oranges, tomatoes, potatoes, nuts, coffee and spices. The varieties used worldwide in agriculture make use of a very limited range of the characteristics possessed by these plants in the wild. If, for example, diseases or changing climates make these plants harder to grow in the future then we may need to make use of some of the original rainforest varieties once again.

Marine ecosystems

These represent a second example of an ecosystem where the activities of human beings may be having a catastrophic effect on the biodiversity of the ecosystem, and hence may eventually destroy its complex ecological balance. In several parts of the world, the population of certain fish species has declined to such low levels that it is doubtful whether they will ever return – for example cod in the waters off Newfoundland. Climate change may be making the situation worse – for example warming ocean waters may be killing the algae in coral reefs, thus removing one of the key producer organisms.

Questions

1 The rainforest has a very high biodiversity.
   (a) Explain what this means.
   (b) Give three reasons for this high biodiversity.
   (c) Suggest one ecosystem, or biome, where the biodiversity might be expected to be much lower.

2 (a) Describe and explain two reasons why many people oppose the destruction of the rainforest.
   (b) Suggest why, despite opposition, rainforests are still being destroyed.

3 Although in some parts of the world only small sections of the rainforest have been cut down – often narrow 'corridors' to allow the building of roads – the ecosystems in these rainforests have been very seriously affected already. Suggest reasons why.

Applications: medicines from the rainforest

1 *Vinblastine* and *Vincristine*, extracted from the rainforest plant periwinkle, are amongst the world's most powerful anticancer drugs. They have dramatically increased the survival rate for acute childhood leukaemia and Hodgkin's disease.
2 It has been estimated that as much as a quarter of all active pharmaceutical ingredients are derived from rainforest sources. But less than 1% of the tropical trees and plants in these forests have been tested by scientists – who knows what other powerful drugs remain to be discovered?
3 The US National Cancer Institute has identified 3000 plants that are active against cancer cells. 70% of these plants are found in the rainforest. About 25% of the active ingredients in today's cancer-fighting drugs come from organisms found only in the rainforest.

Activity

Research a marine ecosystem where biodiversity is threatened – for example coral reefs or the fishing grounds of the North Sea. Find out what strategies are used to try to preserve the ecosystems and how effective these are.

1.2 (6) Natural selection and the development of new species

Genes

Lengths of DNA present in the nucleus of a cell which control the characteristics of organisms. Genes are passed on to the offspring of organisms when they reproduce.

Darwin's finches. A group of birds that he thought were very different but turned out to be closely related. The differences were due to adaptation to eating different foods.

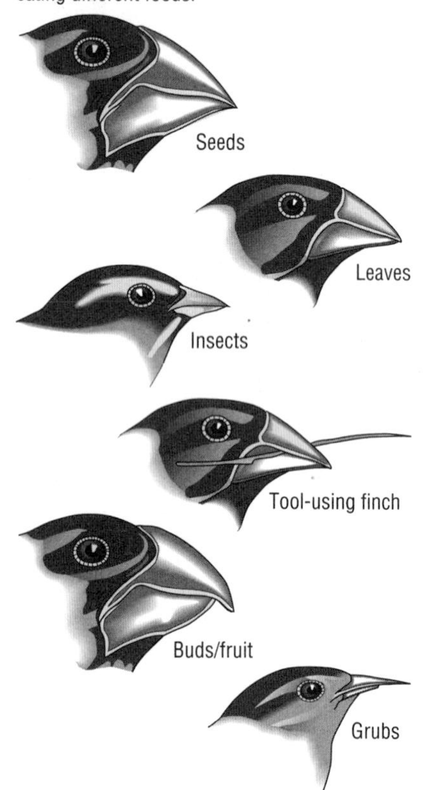

Figure 1 Darwin's sketch showing variation in beak type of finches on the Galapagos Islands

Explaining biodiversity

The rainforest is high in biodiversity because different organisms are adapted to survive in the many different niches in the rainforest. These adaptations may be structural – such as birds with longer beaks, frogs with poisonous skins and plants which root themselves onto tree trunks – or behavioural – for example hunting or feeding only at night. Significant structural or behavioural adaptations result in the formation of a new species.

Natural selection and evolution

The development of these adaptations, and hence of new species, was explained by the theory of *evolution by natural selection.* This was proposed in the nineteenth century by Charles Darwin and Alfred Wallace.

It relies on the fact that within a population there will be variation in characteristics – for example the beak length of a particular type of bird. We now know that this variation is due to the presence of mutations in genes, or simply to the range of ways in which genes have combined together.

The other key point is that it is impossible for all the members of a population to survive long enough to breed. There will always be competition for food, water, sunlight or nutrients as well as the need to avoid predators and to attract mates.

Only individuals that possess the most favourable characteristics will survive to reproduce. These characteristics will then be passed on to the next generation (in the form of genes) – this is the mechanism of adaptation by natural selection.

Natural selection will lead to the evolution of a new species if this process can be repeated many times over a long period of time. Natural selection is described as a *cumulative* effect.

Adaptation

The development of a characteristic which increases the ability of an organism to survive (in a particular niche of an ecosystem).

Speciation

The formation of a new species by cumulative natural selection over a long period of time.

Activity

Because natural selection is a simple mechanism, it is possible to simulate it mathematically – this is known as the 'blind watchmaker' model. By running such a simulation you can discover for yourself how random combinations of genes can produce variety in a population and how selection can cause a life-form to evolve. Search for one on the Internet using 'blind watchmaker + biomorph'.

Two other factors are thought to encourage the process of evolution:

- rapid changes in the environment (such as climate change) which will cause increased competition for survival
- populations that become physically separated from each other (known as geographical isolation) – this means that they cannot interbreed and so the favourable genes become increasingly common in the population.

Rainforest evolution – butterflies

There are at least 7500 different species of butterfly in the Amazon rainforest. Some of these appear to have evolved surprisingly rapidly producing a range of species with different markings. These markings may produce an advantage by camouflaging the butterfly.

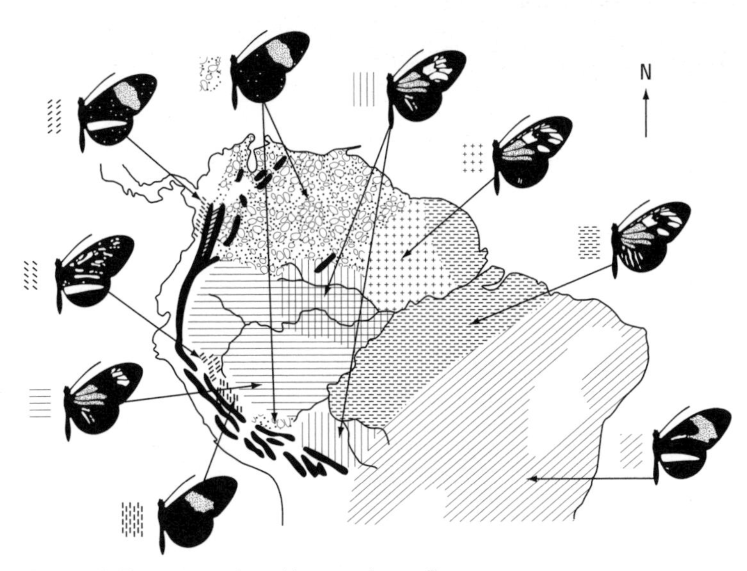

Figure 2 The range of markings on butterflies

One suggestion for the rapid evolution of these butterflies is that it was a result of climate change. This may have made the Amazon rainforest drier and changed the foliage on which the butterflies feed – only the best camouflaged survived. In addition, it is thought that the rainforest may have become separated into several 'islands' surrounded by more open savannah, which is not a suitable habitat for these butterflies – populations of butterflies became isolated in these islands. The combination of environmental change and this *geographical isolation* produced rapid evolution and speciation.

Other scientists disagree about the rainforest island theory. They suggest that once different markings appeared in two populations, then butterflies from these two populations would be unlikely to choose each other as mates. The populations are then said to be in a state of *reproductive isolation*. This would also have encouraged evolution to occur.

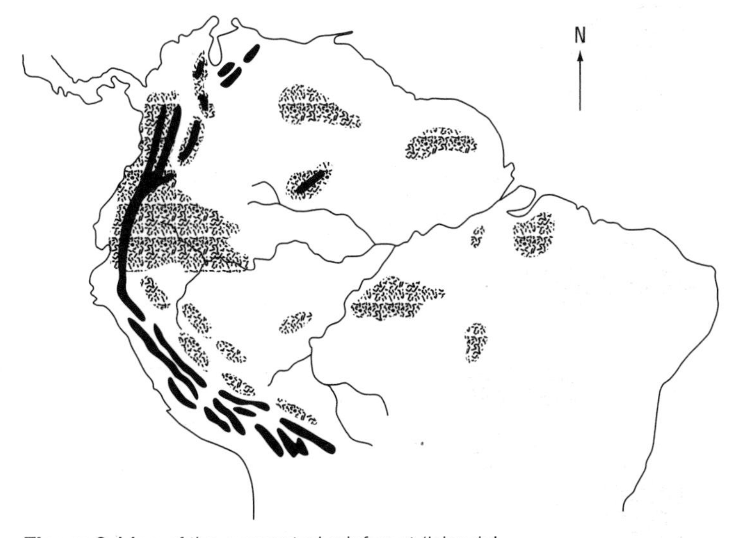

Figure 3 Map of the suggested rainforest 'islands'

Questions

1. Natural selection is described as a 'cumulative process'. What does this mean?
2. Members of a species are said to be 'in competition with each other'. What does this mean?
3. Explain what is meant by a 'favourable characteristic'. Why is this more likely to be passed on to the next generation than one which is not favourable?
4. Darwin called his theory 'natural selection' because he thought about the way in which wild wolves had been gradually transformed into domestic dogs by selective breeding.
   (a) Outline how this has been used to produce new breeds of dogs with very long coats – for example Afghan hounds.
   (b) Do you think that a new breed represents a new species? Justify your answer.
5. It is thought that evolution by natural selection does not proceed at a steady rate but occurs very rapidly in response to certain events – for example rapid climate changes such as those experienced at the beginning and end of ice ages. Describe reasons why climate change might cause rapid evolution.
6. It is said that human beings have stopped evolving. Why might this have happened? Do you agree with the statement?

1.2 ⑦ The nitrogen cycle

Nutrients

Plants are the producers of most ecosystems – they trap light energy and use this energy to create biomass from carbon dioxide and water.

In addition to light energy, carbon dioxide and water, plants also need substances called **nutrients** to develop and grow healthily. These nutrients must provide certain chemical **elements** – for example nitrogen, phosphorus and calcium.

Nutrient element

An element (type of atom) required for the healthy development of organisms.

Nutrient cycles

Nutrients are required by all living organisms. So, in the same way in which energy (in the form of biomass) is passed on from one organism to another, nutrients also pass through an ecosystem. Nutrient cycles, such as the nitrogen cycle, show how this happens.

- Cycles consist of stores (reservoirs) of the nutrient element.
- There is a flow of nutrient element (a flux) between different reservoirs in the cycle.
- The nutrient element is converted into different chemical substances as it passes around the cycle.

Examiner tip

You will not be expected to draw the whole nitrogen cycle from memory. However, you may be expected to explain or describe how the cycle works, or to add information to an incomplete cycle.

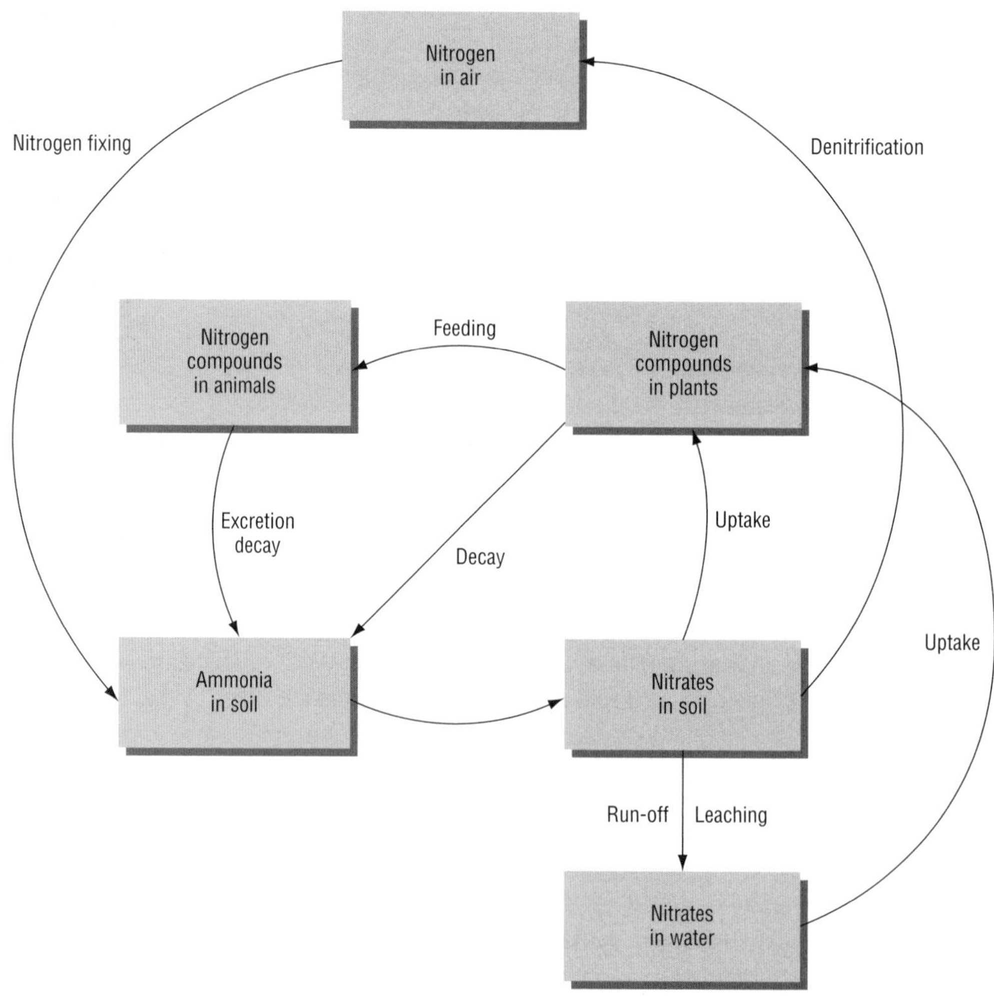

Figure 1 The nitrogen cycle

Inputs and outputs

Nutrient cycles are a useful way of describing the processes in a single ecosystem – for example in a field of crops. This is an artificially maintained ecosystem, so to ensure that it remains sustainable the input and output must remain balanced. In natural ecosystems, processes of negative feedback ensure that the ecosystem remains in a steady state.

Steady state

A situation in which the level of something (e.g. the amount of nutrient in a reservoir) remains constant. The input of nutrient into the reservoir must be equal to the output.

Negative feedback

This describes a process which opposes or reverses a change which occurs in a system. It can restore the system to steady state – where input is equal to output.

In the example of a field of crops, natural inputs will include:

- the effect of nitrogen-fixing bacteria which convert atmospheric nitrogen into ammonium compounds (which contain the NH_4^+ ion) – these bacteria live naturally in soil but some plants (peas, beans and clover) have nodules on their roots which contain vast numbers of them
- nitrogen fixing caused by lightning, which converts atmospheric nitrogen into nitrogen oxides (NO_x).

Natural outputs include:

- leaching – nitrogen compounds dissolve in rainwater percolating through the soil; the nitrogen compounds eventually end up in aquifers (groundwater)
- run-off – nitrogen compounds on the surface dissolve in rainwater during heavy rain and wash into streams, rivers and lakes
- denitrifying bacteria – convert nitrate ions (NO_3^-) into atmospheric nitrogen.

There is an extra output in a farmed-field ecosystem because the crops are removed by harvesting. To ensure a steady state, this output must be matched by an input. Farmers can add fertilisers (see spread 1.2.9 *Fertiliser manufacture and eutrophication*) or regularly grow peas, clover or beans in the field. Higher fertiliser levels will cause a new steady state to be produced with a higher nitrogen content in the soil. This will encourage more plant growth.

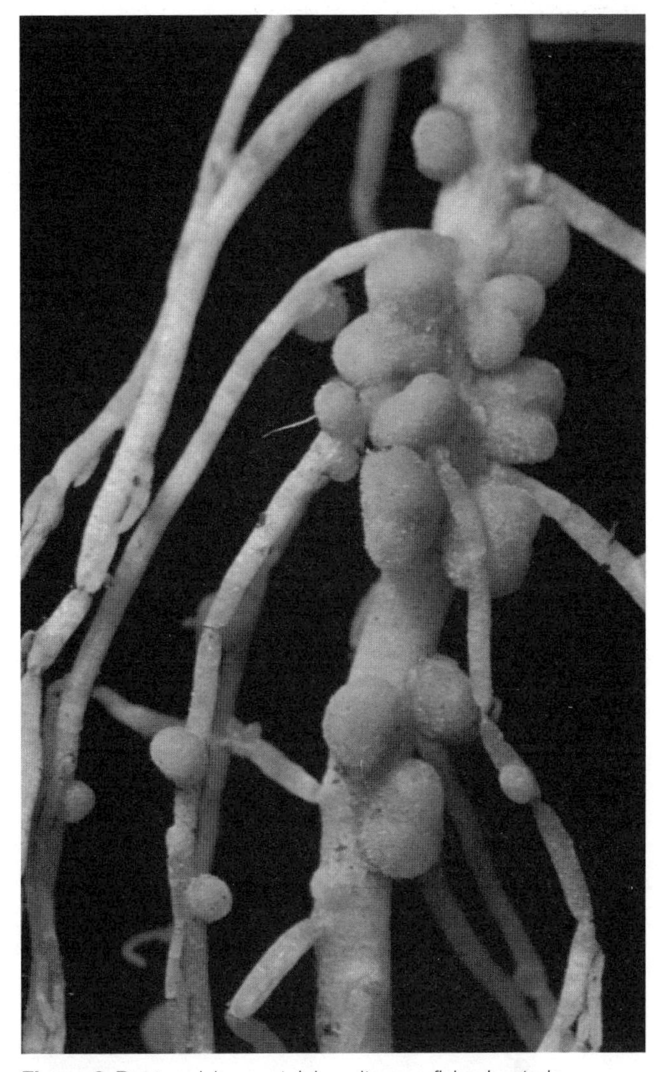

Figure 2 Root nodules containing nitrogen-fixing bacteria

Uptake and return

Plants can absorb nitrate ions (NO_3^-) through their roots and then use the nitrates in their metabolic processes – to produce proteins for example. Other forms of nitrogen (for example ammonium ions) can be converted into nitrates by nitrifying bacteria.

Dead plant material decays back into the soil by the action of decomposers (bacteria and fungi), releasing ammonium ions and nitrates. Some plant material is eaten by animals – this is returned to the soil in faeces or by decomposition when the animal dies.

Other ecosystems and other nutrient cycles

This general description of the nitrogen cycle can be adapted to describe nutrient cycling in different ecosystems – for example in ponds, rivers and the sea, where run-off and leaching may be *inputs* into the ecosystem rather than outputs. In these situations, human activities, such as the overuse of fertilisers or the release of sewage, may be causing damaging effects on the nutrient cycle. This is considered in spread 1.2.9 *Fertiliser manufacture and eutrophication*. There are also nutrient cycles for other elements as well – for example the phosphorus or calcium cycle.

Examiner tip

You will not be expected to remember details of cycles other than the nitrogen cycle, but you may be given information about them which will enable you to construct them for yourselves.

Questions

1 Give two examples of:
   (a) inputs into a nitrogen cycle
   (b) reservoirs of nitrogen
   (c) fluxes between the reservoirs
   (d) outputs from a nitrogen cycle.
2 (a) Explain the difference between 'run-off' 'and 'leaching'.
   (b) Explain the difference between nitrifying bacteria and denitrifying bacteria.

1.2 (8) Nutrient fluxes and feedback

Analysing nutrient fluxes

A nitrogen cycle in an isolated ecosystem can be represented by a simple diagram such as that shown in Figure 1.

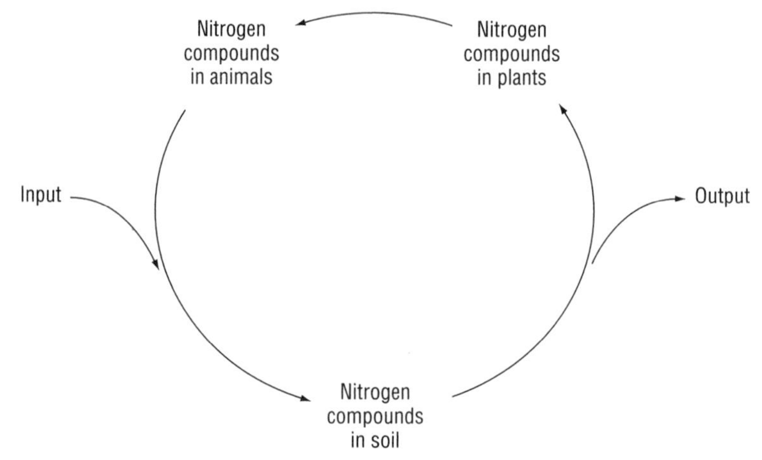

Figure 1 Diagram showing input, output and a simplified cycle

Examiner tip

You will need to recognise whether a system is in a steady state by checking that the input and output are equal and that the fluxes in and out of each reservoir are equal.

You may be given data for only some of the inputs, outputs and fluxes. If the system is in a steady state you can then work out the missing values.

If the ecosystem is in a steady state you can conclude two things about the system:

- the input and the output are equal
- the amount of nitrogen in each reservoir remains equal – this means that the fluxes in and out of each reservoir must also balance.

Feedback

You came across negative feedback in spread 1.2.7. Negative feedback loops keep ecosystems in a steady state. For example, if extra fertiliser is applied then this will increase the amount of nitrogen in the soil reservoir. A negative feedback process will occur to reduce the amount of nitrogen in the soil and produce a new steady state. In this case, the extra nitrogen will promote more plant growth, resulting in greater uptake from the soil. The higher concentration of nitrogen in the soil also means that the rate of leaching will increase, which will also reduce the amount of nitrogen in the soil.

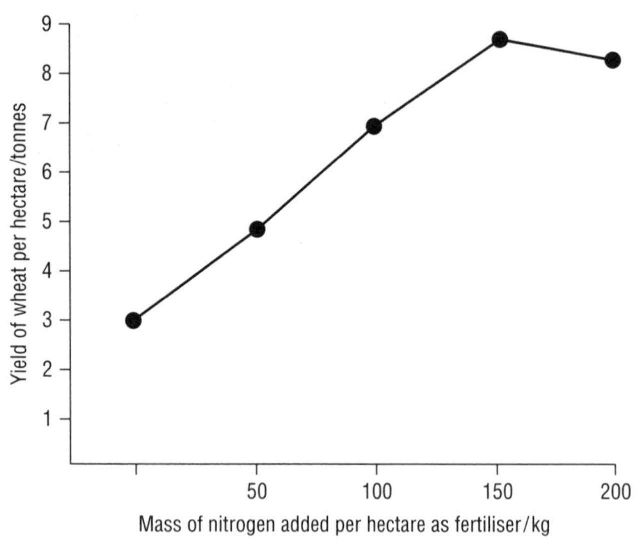

Figure 2 Graph showing yield of wheat against mass of nitrogen added as fertiliser

The effect of fertilisers

Much research is carried out to assess the effect of increasing nutrient levels on plant productivity (normally in terms of crop yields). Results often show that there is a limit beyond which increased use of fertilisers does not produce any further increase in yield.

Environmental and economic factors will mean that the optimum amount of fertiliser may be rather lower than that suggested by the graph.

Positive feedback

Positive feedback loops are much less common in biological systems than negative feedback. They tend to make changes occur at ever-increasing speeds – they amplify the effect of change, which is normally a bad thing. Most biological systems – ecosystems or organisms – will be sustainable only in a very narrow range of conditions. Anything which causes these conditions to change too much will cause the ecosystem to collapse or the organism to die.

Positive feedback

A process which increases the effect of a change which occurs in a system. A small initial change in the system will result in a very large final change.

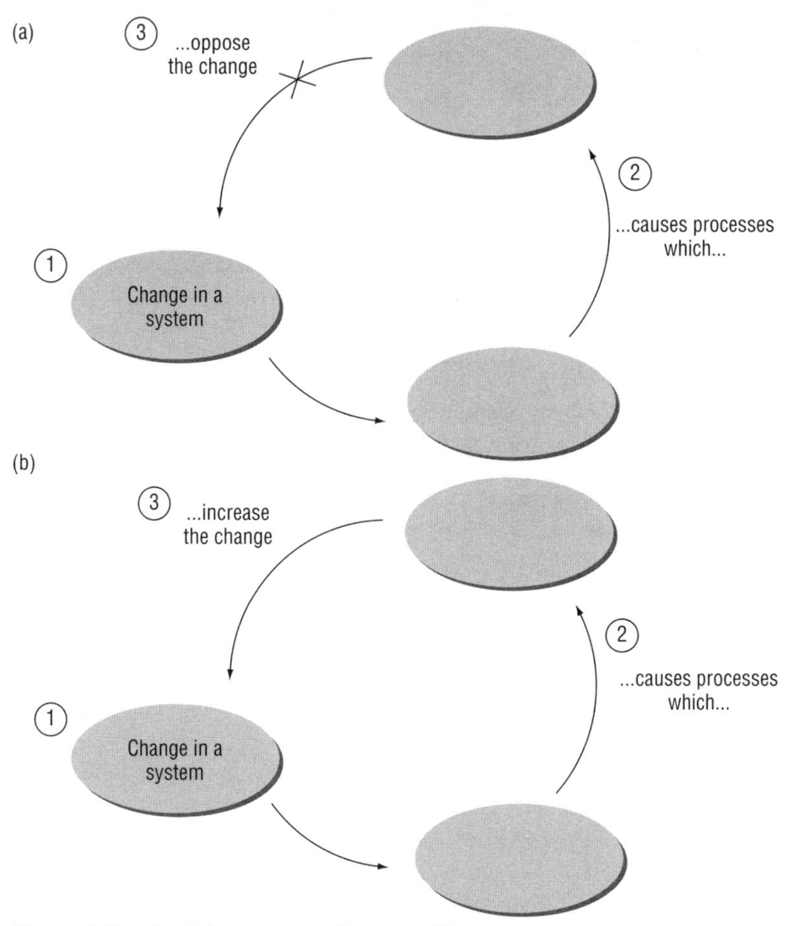

Figure 3 Feedback loops: **a** negative; **b** positive

Although positive feedback is rarely seen in biological systems, there are many important examples of positive feedback in the natural world, particularly in the way the Earth may respond to climate change.

Questions

1 All ecosystems rely on a number of nutrient cycles.
- **(a) (i)** What is meant by the term 'nutrient'?
  - **(ii)** Name two other elements (other than nitrogen) that are normally classified as nutrients.

2 State the names of natural processes by which:
- **(a)** nitrogen is input into an ecosystem from the atmosphere
- **(b)** nitrogen is lost from the soil.

3 An ecosystem and all the reservoirs of nitrogen in it are said to be in a steady state.
- **(a)** What is meant by 'steady state'?
- **(b)** The input of nitrogen is 2.5 kg ha^{-1} yr^{-1}. Within the ecosystem 30.2 kg ha^{-1} yr^{-1} is taken up by the roots of plants. Some of this is returned to the soil by decay, and some is passed on to animals by feeding. 4.0 kg ha^{-1} yr^{-1} is returned from animals to the soil by decay and excretion. Draw a diagram to summarise this information.
- **(c)** Assuming that the ecosystem is in a steady state, calculate values for the output and the two missing values of the fluxes between reservoirs in the ecosystem.

4 Nitrogen is returned to the soil from trees mainly by the process of leaf fall.
- **(a)** What further process must occur before the nitrogen is converted into nitrates?
- **(b)** Negative feedback occurs in this system to control the level of nitrates in the soil.
  - **(i)** Explain what is meant by the term 'negative feedback'.
  - **(ii)** Describe what would happen to the fluxes shown in your diagram in **3(b)** if the level of nitrates suddenly rose – if some artificial fertiliser had been added for example.

Applications: positive feedback in climate change

One effect of global warming will be that some of the permanent ice and snow trapped in glaciers or ice-caps will melt. This will expose the bare rock underneath. Bare rock absorbs energy from sunlight much more effectively than the lighter ice and snow, and so the Earth's surface heats up more rapidly, causing even more ice and snow to melt. This could cause a dramatic increase in the rate of global warming.

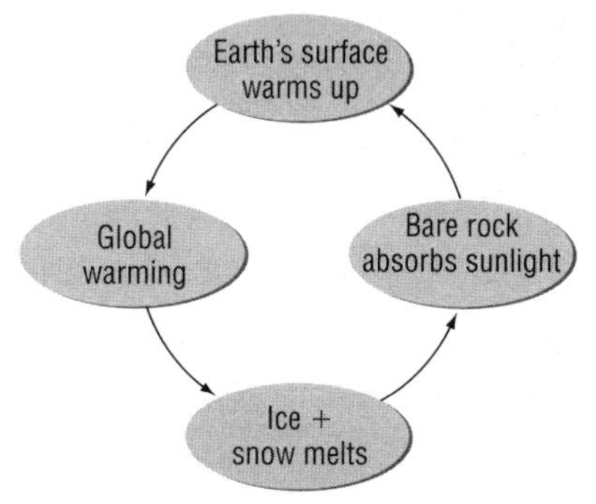

Figure 4 Diagram to show the reflection of light by ice-caps, the absorption of light by rock and the increased melting of the ice-caps

Activity

You can find out about more feedback loops which may affect the rate of global warming. Try searching the Internet for 'feedback + global warming'.

1.2 (9) Fertiliser manufacture and eutrophication

How science works

In this spread you will consider the environmental implications of industrial nitrogen fixing (HSW 6a) and of the eutrophication caused by the application of nitrogen-based fertilisers (HSW 6a)

Nitrogen fixing and the Haber process

Nitrogen fixing is an extremely difficult process because nitrogen gas in the atmosphere is so inert – it will not react with other substances easily. This is because the bonds holding the atoms together in a nitrogen molecule are very strong.

In 1909 Fritz Haber developed a process that allowed nitrogen from the air and hydrogen (which can be obtained from natural gas) to react to produce ammonia:

nitrogen + hydrogen → ammonia

$$N_2 + 3H_2 \rightleftharpoons 2NH_3$$

N≡N

Figure 1 A nitrogen molecule – the very strong triple bond makes it very difficult to pull the nitrogen atoms apart

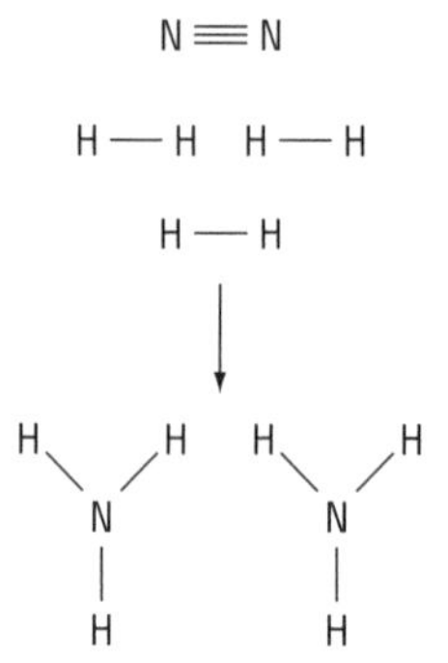

N≡N

H—H H—H

H—H

↓

NH_3 NH_3

Figure 2 What happens to the atoms in the Haber process

The reaction requires specific conditions to produce economic amounts of ammonia:

- a fairly high temperature (400–500°C)
- a high pressure (around 100 times atmospheric pressure – written as 100 atm)
- a catalyst of iron.

Although the raw materials for making ammonia are relatively cheap, a great deal of energy must be used to produce these conditions. This energy is obtained from the burning of fossil fuels, releasing carbon dioxide and other pollutants. Approximately 1% of the total mass of fossil fuels burned per year is used in fertiliser manufacture.

Ammonia can be converted into nitrates in a complex series of steps:

ammonia + oxygen → nitrates + water

A common fertiliser is ammonium nitrate, which contains ammonium ions (NH_4^+) and nitrate ions (NO_3^-) which are both sources of nitrogen for the soil.

Examiner tip

You do not need to know the chemical formulae of the substances in the process but you should be able to describe what happens using a word equation and discuss the significance of the conditions.

Catalyst

A substance which speeds up the rate of a chemical reaction but is not used up – and can therefore be re-used.

Eutrophication

Apart from the issue of energy use in manufacturing fertilisers, there is another major environmental problem caused by the overuse of fertilisers.

Nitrate-based fertilisers are soluble – run-off (or leaching) causes them to be lost from the soil by dissolving in rainwater and they end up in rivers, lakes and seas, or in the groundwater which may eventually feed into them. This causes a process known as **eutrophication**. Phosphate-based fertilisers have the same effect, as do the phosphates from detergents or animal waste.

Eutrophication disrupts the natural nutrient cycles in water-based ecosystems. An increase in nutrient causes rapid population growth of algae and other water-based plants.

As the large population eventually starts to die and decay at an increased rate, the decomposer organisms begin to use up the oxygen in the system. This may be made worse if the algae float on the surface, blocking out sunlight and preventing photosynthesis below. Most of the organisms in the ecosystem, such as fish, cannot survive in the low oxygen level and die.

Eutrophication

An increase in the level of nutrients, normally seen in aquatic ecosystems.

In some cases, fish which are tolerant of low oxygen levels, such as roach, will replace other species which require higher oxygen levels, such as salmon and trout. So even if the ecosystem remains a functioning one, biodiversity may be greatly reduced.

It is important to realise that eutrophication, and the change in populations which results, is a natural process but it will normally take place over thousands or millions of years. Artificial eutrophication can occur in just one season.

Nitrates in water supply

Some of the nitrates lost from the soil through leaching end up in the underground aquifers which are the main source of drinking water for human populations. In some cases the nitrates (NO_3^-) can be reduced to nitrites (NO_2^-) which are toxic as they can affect the ability of the haemoglobin in the blood to carry oxygen.

Levels of nitrate of 50 mg dm^{-3} and below are regarded as safe – water is tested regularly to ensure that it is below this level. No cases of illness due to nitrite poisoning have been recorded in the UK since the 1950s.

Questions

1 (a) Explain why the production of ammonia from nitrogen and hydrogen is important for food production.
   (b) What are the raw materials from which nitrogen and hydrogen are obtained?
   (c) What conditions are used to ensure that ammonia is produced economically?
   (d) Suggest why it is difficult to 'fix' nitrogen without these specialised conditions.

2 (a) Describe how the process of eutrophication can lead to changes in aquatic ecosystems.
   (b) State one other problem which can be caused by the use of nitrogen-based fertilisers on soil.

1.2 Module 2 – Summary

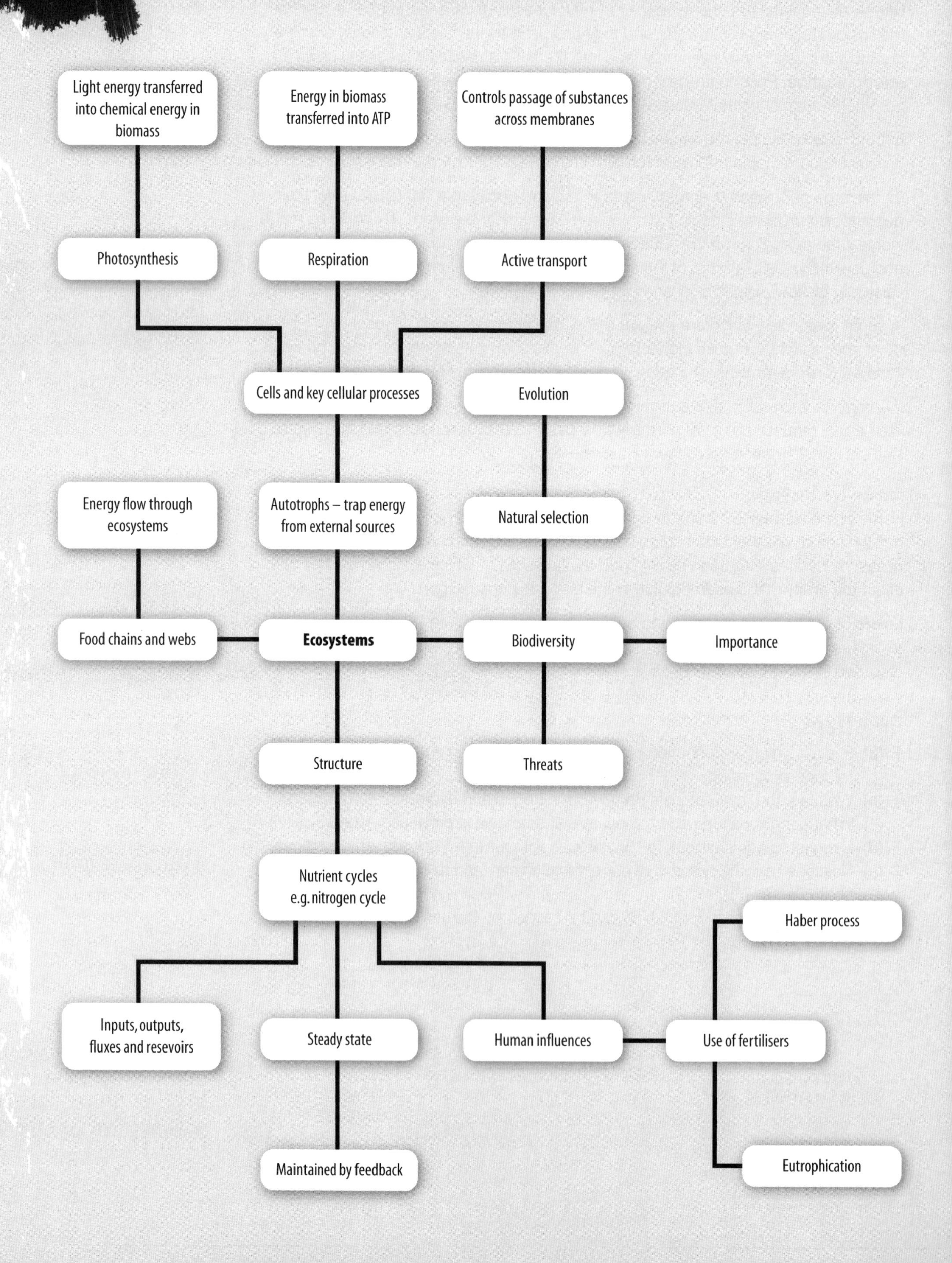

Practice questions

Low demand questions

These are the sort of questions that test your knowledge and understanding at E and E/U level.

1 Match up the following names of molecules with their correct function

Chlorophyll	Component of the cell membrane
ATP	Product of photosynthesis; acts as a store of chemical energy
Glucose	Product of respiration; acts as a mobile energy store
Carbon dioxide	Absorbs light energy and transfers it to chemical energy
Lipid	Used as the source of carbon atoms in the manufacture of biomass by photosynthesis

2 Marine and rainforest ecosystems are examples of ecosystems which have been affected by the actions of humans. In both cases the biodiversity of the ecosystem is under threat.

(a) Describe one way in which human actions are affecting:

(i) rainforest ecosystems

(ii) marine ecosystems.

(b) (i) State the meaning of the term 'biodiversity'.

(ii) Give one reason why maintaining the biodiversity of the rainforest is important for human societies.

Medium demand questions

These are the sort of questions that test your knowledge and understanding at C/D level.

3 In Amazonian rainforest ecosystems, much of the forest floor is covered by dead leaves which fall from trees. These are used by ants as a source of food. The ants are eaten by anteaters, which in turn may be eaten by predators such as jaguars.

(a) Construct a food chain to show the flow of energy through this ecosystem. Use your food chain to explain the term producer.

(b) Very little of the biomass in the leaves is passed on to the organisms at the top of the food chain. Explain what happens to the energy in this biomass.

4 It is thought that many of the big cat predators – for example jaguars, tigers and ocelots – may have evolved from a single ancestor. Outline how the process of evolution by natural selection may have produced these different species.

High demand questions

These are the sort of questions that test your knowledge at A/B level.

5 (a) Glucose and ATP can both be regarded as stores of chemical energy. Explain how they have different roles in cells.

(b) 1 mole of glucose can release approximately 2800 kJ of energy when it is broken down in anerobic respiration. 30 kJ of energy are required to synthesise 1 mole of ATP.

(i) Calculate the maximum number of moles of ATP which could, in theory, be formed from the breakdown of 1 mole of glucose.

(ii) In practice, only about 30 moles of ATP are actually formed. Calculate the % of energy from glucose transferred to ATP.

6 The amount of nutrients in the soil of an ecosystem tends to remain constant, even if small changes occur in the conditions of the ecosystem. This is due to processes described as negative feedback.

(a) (i) Explain the meaning of the term 'negative feedback'.

(ii) Describe how negative feedback might operate to maintain the amount of nutrients in the soil if small amounts of artificial fertiliser, such as ammonium compounds, are added to the soil.

(b) One key process which occurs during the formation of the ammonium compounds is the reaction of nitrogen with hydrogen to form ammonia.

(i) What conditions are used in industry for this reaction?

(ii) Explain how the manufacture of ammonia can be regarded as environmentally damaging.

(iii) Describe one way in which the use of fertilisers such as ammonium compounds can also cause environmental problems.

UNIT 2

Module 1 Weather, climate and climate change

Introduction

In recent years, we have become all too aware of the destructive and devastating power of the weather. Floods, droughts, hurricanes and heatwaves in many parts of the world – including Britain – have become frequent headline stories in newspapers and television bulletins. You will see how knowledge of the structure of the atmosphere and of the behaviour of gases will help you to make sense of such weather events. You will use a simple model of gases – the kinetic theory – but later you will see how this model needs to be adapted when we consider the behaviour of perhaps the most important molecule on Earth – water. Molecules of water turn out to behave in unexpected ways and this helps to explain why not only the atmosphere but also the ocean currents need to be understood in order to make sense of our weather.

Closely linked to the science of weather is the science of climate – the average weather we experience over a long period. We will look at the evidence which helps us to decide whether the climate is really changing as quickly as many believe, and we will start to explain why that is and whether or not human beings really have any chance of averting catastrophic climate change.

The story of climate change, its causes and possible solutions is a long and complex one, and we will continue to develop this story in Modules 2 and 4.

Module contents

1 Atmospheric circulation and pressure
2 Pressure and kinetic theory
3 Movement of air in the atmosphere
4 Water and covalent bonding
5 The polar water molecule and its unusual properties
6 Ocean circulation
7 The thermohaline circulation
8 Evidence for climate change
9 Models of climate change
10 Analysing climate data

How science works

During this module you will covering some of the aspects of How Science Works. In particular you will be studying material which may be assessed for:

- HSW 5b: Analyse and interpret data to provide evidence, recognising correlations and causal relationships.
- HSW 5c: Evaluate methodology, evidence and data, and resolve conflicting evidence.
- HSW 7a: Appreciate the tentative nature of scientific knowledge.
- HSW 7b: Appreciate the role of the scientific community in validating new knowledge and ensuring integrity.
- HSW 7c: Appreciate the way in which science uses science to inform decision making.

Examples of this material include:

- Examine and evaluate the evidence for climate change and its likely causes (spread 2.1.8).
- Compare and contrast the models of future climate (spread 2.1.9).
- Discuss the range of possible responses to the issue of climate change (spread 2.1.9).

Test yourself

1 Draw a diagram to show the arrangement of molecules in a gas.
2 Explain what is meant by the terms 'pressure' and 'density'.
3 The UK and the Sahara desert in north Africa have very different climates. Describe some of the differences, and try to explain why the two climates are so different.
4 Explain what is meant by *molecule*, *electron*, *nucleus* and *covalent bond*.
5 What gases are thought to be responsible for the greenhouse effect? What processes produce these gases?

2.1 ① Atmospheric circulation and pressure

Troposphere

The lowest layer in the Earth's atmosphere extending up to around 15 km from the Earth's surface.

Stratosphere

A layer in the Earth's atmosphere between 15 and 50 km from the Earth's surface.

Tropopause

The boundary between the troposphere and the stratosphere.

The structure of the atmosphere

The Earth's atmosphere can be thought of as a series of layers. In this course you will be looking at processes in the **troposphere** and **stratosphere**, which are separated by the tropopause.

The lower atmosphere (**troposphere**) consists of moving air masses. Satellite images can show that there are specific regions of the Earth in which air masses tend to be either moist or dry.

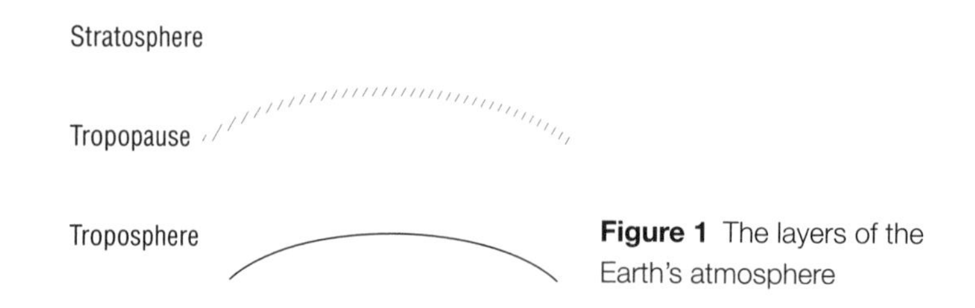

Figure 1 The layers of the Earth's atmosphere

The link between air mass type and air pressure

Air masses can be described as being **high pressure** or **low pressure**. You often see these terms used in weather maps in newspapers and on the television news. The pressure in these maps is measured in millibars (mB). How gases exert pressure is explained in spread 2.1.2 *Pressure and kinetic theory*.

The average atmospheric pressure is around 1012 mB. In a region of high pressure, the typical pressure might be 1024 mB; in a region of low pressure it might be 1000 mB. A typical distribution of high and low pressure areas over the Earth's surface is shown in Figure 2.

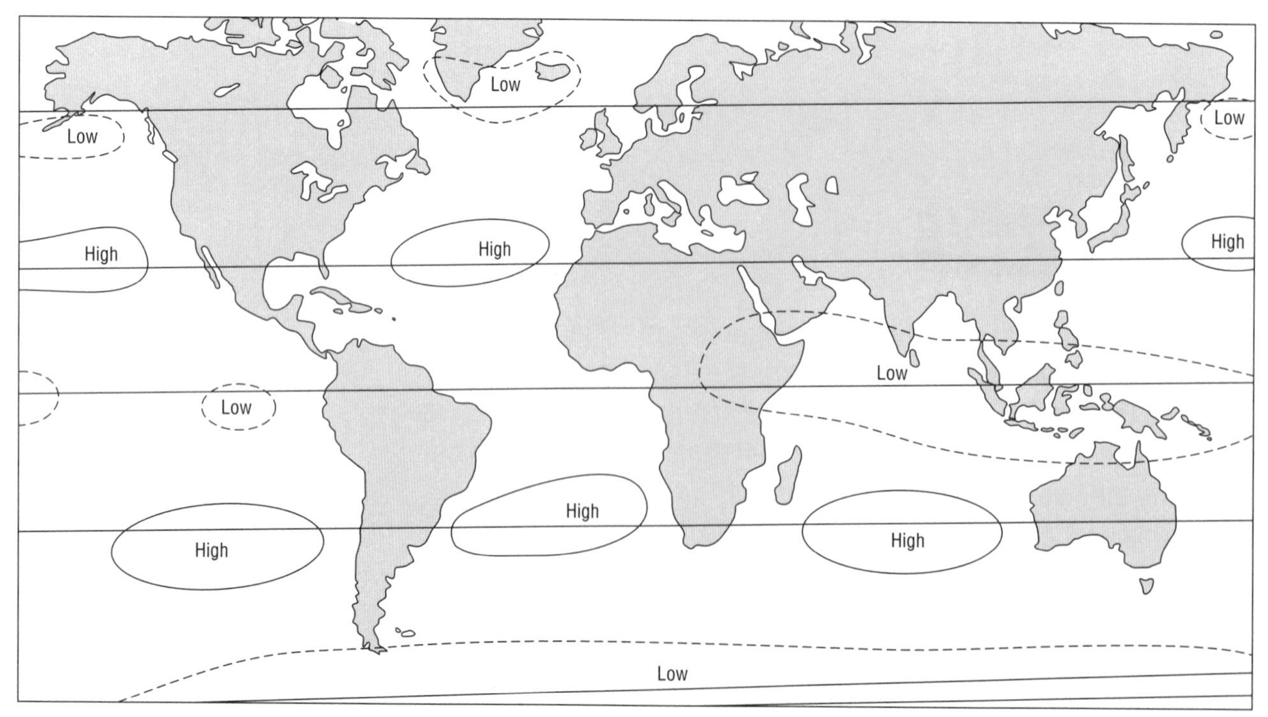

Figure 2 Map showing distribution of high and low pressure

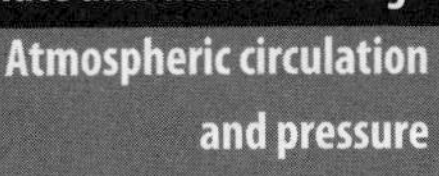

Figure 3 Satellite image showing the zones of dry and moist air – moist air is indicated by the presence of clouds (white)

Activity

There are numerous websites which show up-to-date images, often animated, of the cloud patterns over the globe. You can use these images to look at how the zones of dry and moist air are distributed at different times of the year.

Areas of high pressure are generally associated with dry climates; areas of low pressure are associated with wet climates.

The poles do not show up easily on some satellite images – and cannot be seen in Figure 2 – but they have permanent high pressure air masses.

The temperature of air masses

The surface of the Earth is warmed as it absorbs electromagnetic radiation from the Sun. The air above the surface is heated by contact with this warm surface. The temperature of the Earth's surface (and hence the air masses above it) decreases as the distance from the equator increases. This can be explained by considering the intensity of the radiation received at various points on the Earth's surface, as shown schematically in Figure 4.

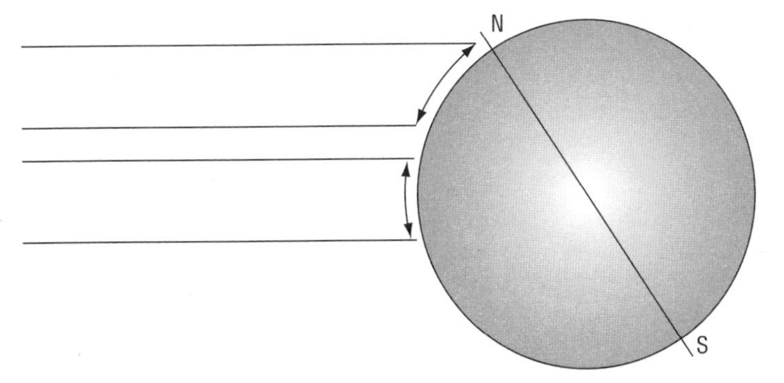

Figure 4 Solar radiation is spread out over a wider area close to the poles

Climate zones

The Earth can be divided into several broad climate zones depending on the typical pressure and temperature of the air mass present – these are shown in Figure 5.

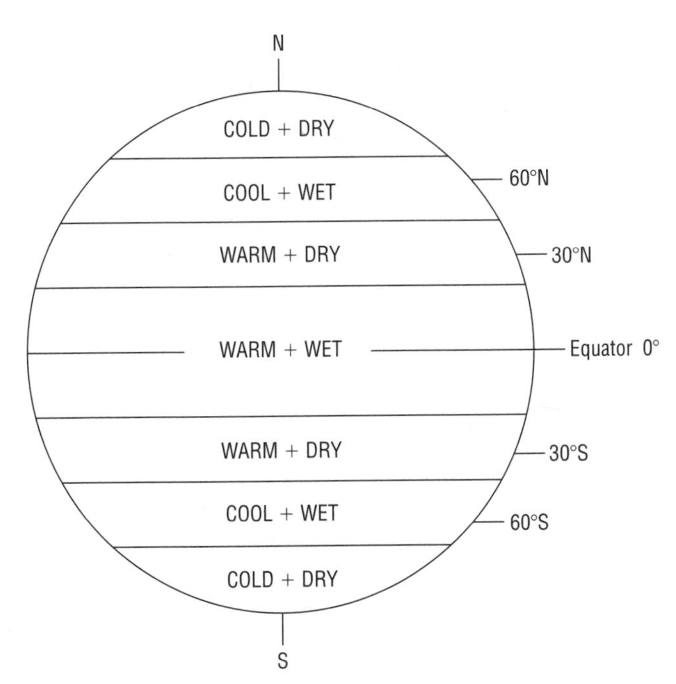

Figure 5 Climate zones at different latitudes

Questions

1. Give two factors that affect the climate of a region.
2. Put the following features in order of height above the Earth's surface: stratosphere, troposphere, tropopause.
3. The Sahara desert has a hot, dry climate.
   - **(a)** Predict a typical value for the air pressure at the surface of the Sahara desert.
   - **(b)** **(i)** Suggest at what latitude the Sahara desert is found.
     - **(ii)** Use this to explain why the climate of the Sahara is hot.

Pressure and kinetic theory

The molecular kinetic theory of matter

All matter is made up of *particles* which are *moving continuously*. The particles could be molecules, atoms or ions; the particles could be moving from place to place or simply vibrating about a fixed point.

Pressure

The force acting on a fixed area. For example, the force exerted on a 1 m^2 area – in this case it would have the units $N\,m^{-2}$ – also known as pascals (Pa).

Pressure

Gases (and liquids) exert a pressure on the walls of the container that holds them. This happens because when the particles collide with the wall they exert a force on the wall as they bounce back off it. Pressure can be calculated from the *force* and the *area* on which that force is acting:

$$\text{pressure} = \frac{\text{force}}{\text{area}}$$

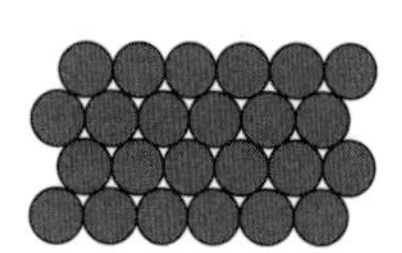

(a) **Solid**
Molecules close together, in a regular pattern, oscillating randomly about fixed positions

(b) **Liquid**
Molecules close together, in a disordered arrangement, moving around randomly, jostling one another

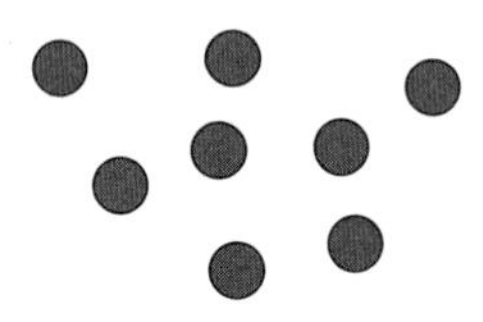

(b) **Gas**
Molecules separated, in a disorderly arrangement, moving around randomly, continuously colliding

Figure 1a Solid – particles arranged regularly but vibrating; **b** liquid – particles arranged randomly and moving past each other; **c** gas – particles far apart and moving randomly

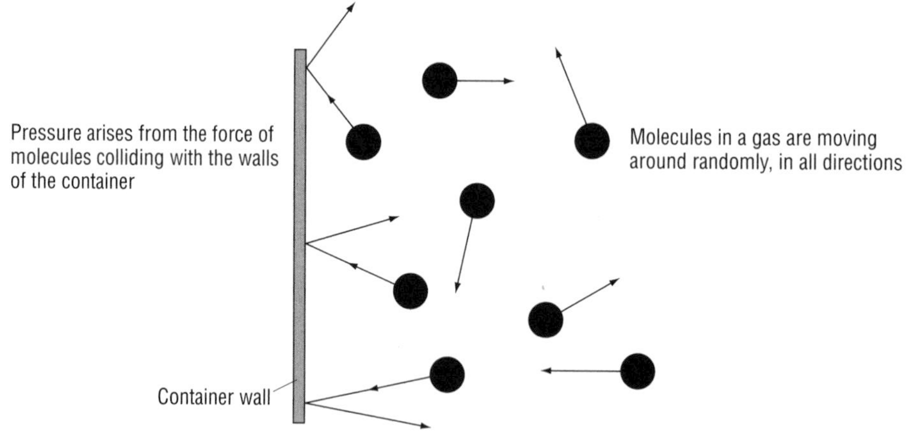

Figure 2 Molecules in a gas colliding elastically with a surface

The effect of changing conditions

We can use the ideal gas laws to predict what happens when conditions are changed.

When the temperature of a gas is increased, the molecules gain more kinetic energy and move faster. This means that when they collide with the walls of their container they exert a greater force, increasing the pressure. This pattern is summarised by the pressure law– at constant volume, the pressure is directly proportional to temperature. This proportional relationship means that if the temperature is doubled then the pressure will also be doubled. However, this relationship is true only if the volume remains constant.

If the same number of molecules are squeezed into a much smaller volume then they will collide with the walls of their container more often – this means that the pressure will increase. The pattern is summarised by Boyle's law – at constant temperature, the

Examiner tip

When using the ideal gas laws, the temperature *must* be given in kelvin. The zero of the kelvin scale is at −273 °C, which means that 0 °C is the same as 273 K – to convert Celsius to kelvin you add 273.

pressure is inversely proportional to the volume. This inverse relationship means that if the volume is doubled then the pressure will halve. Note that this relationship is only true if the temperature remains constant.

The final ideal gas law is known as Charles' law – at constant pressure, the volume is directly proportional to the temperature.

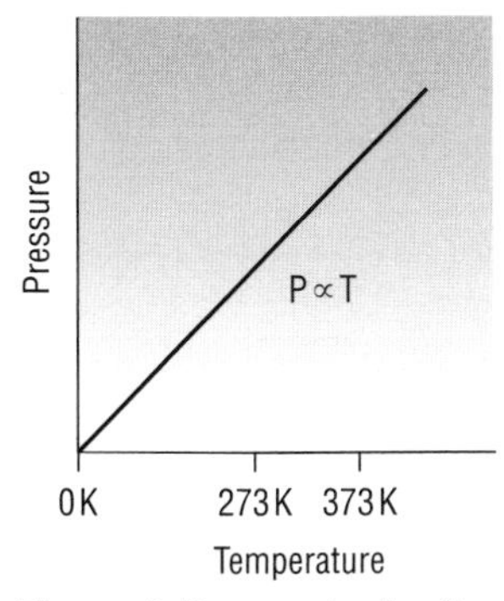

Figure 3 Pressure is directly proportional to temperature

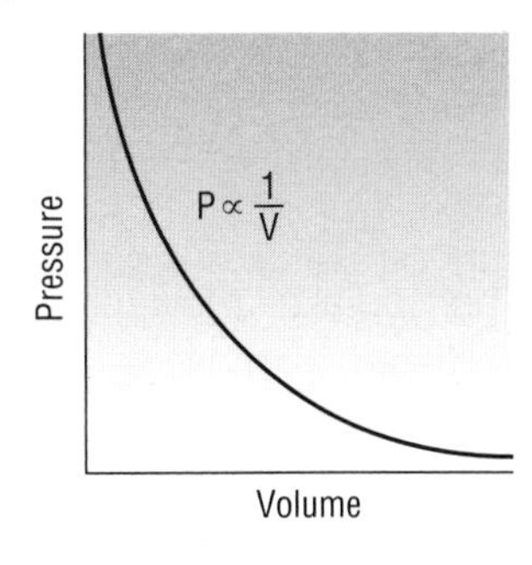

Figure 4 Boyle's law – pressure is inversely proportional to volume

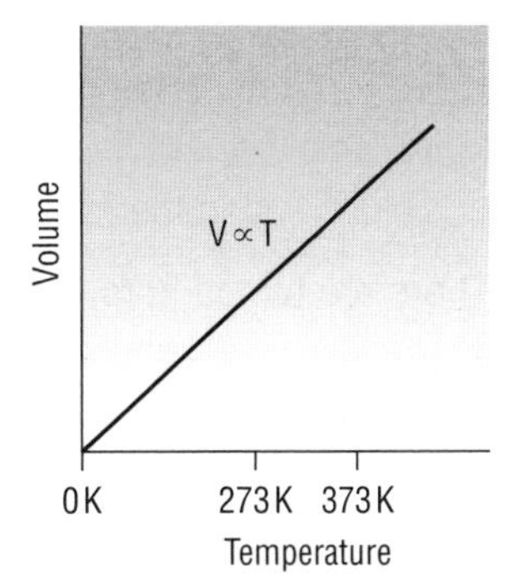

Figure 5 Charles' law – volume is directly proportional to temperature

Changing conditions in the atmosphere

Boyles's law and the pressure law are helpful models in understanding simple situations, such as in laboratory experiments. In the atmosphere, the situation is more complicated – all three factors may be changing at once. Although we would find it difficult to predict these changes by calculation, we can use ideas about kinetic theory to help us make sense of the changes.

Imagine a hot bubble of air expanding in the atmosphere. It has to push the surrounding atmosphere out of the way – see Figure 6a. This requires energy – we say that work has been done on the surrounding air. The energy comes from the molecules in the bubble of air. So these molecules lose energy and their temperature drops – so expanding air cools down.

A similar thing happens when a bubble of air contracts – see Figure 6b. The surrounding atmosphere pushes the molecules closer together – the atmosphere does work on the molecules in the bubble, and these molecules gain energy. So the gas heats up.

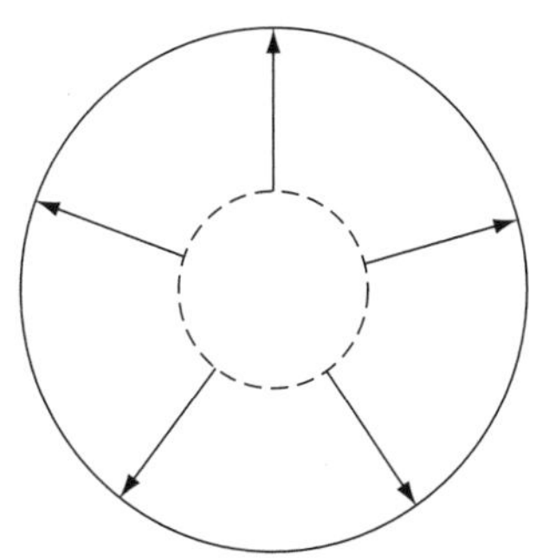

(a) The molecules of air inside the bubble do work as they push on the atmosphere. They lose energy and cool down

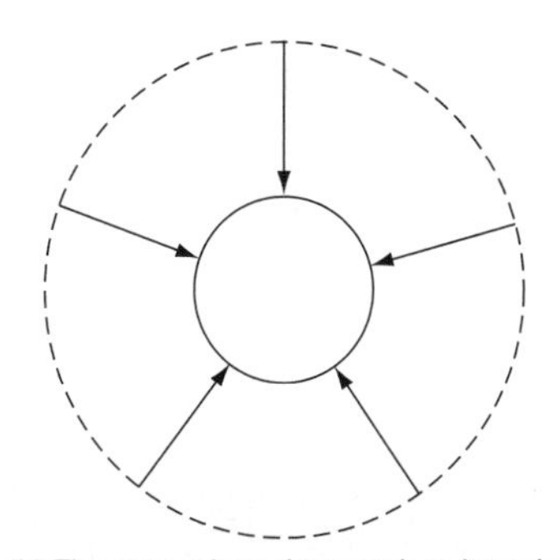

(b) The atmosphere does work as it pushes the molecules in the bubble closer together. They gain energy and heat up

Figure 6 Bubble of air **a** expanding; **b** contracting

Questions

1 Calculate the pressure (in $N\,m^{-2}$) in the following situations:
   (a) a force of 10 N is applied over an area of $2\,m^2$
   (b) a force of 1 N is applied over an area of $0.001\,m^2$
   (c) a force of 5000 N is applied over an area of $500\,m^2$.
2 Calculate the force applied if:
   (a) there is a pressure of $2000\,N\,m^{-2}$ on an area of $8\,m^2$
   (b) there is a pressure of $2\,N\,m^{-2}$ on an area of $0.000\,08\,m^2$
   (c) there is a pressure of $4.0 \times 10^6\,N\,m^{-2}$ on an area of $4.0 \times 10^{-4}\,m^2$.
3 If $5\,dm^3$ of gas is heated from 200 to 600 K, what will the new volume be? Assume that the pressure is constant.
4 If the pressure on $10\,dm^3$ of gas is increased from 100 000 to 200 000 $N\,m^{-2}$, what will the new volume be? Assume that temperature remains constant.

2.1 (3) Movement of air in the atmosphere

Horizontal air movement

Movement of air can occur horizontally across the Earth's surface or vertically, rising or sinking within the atmosphere.

There are two factors that affect horizontal air movement:

- Pressure differences between different air masses – air moves from regions of high pressure to regions of low pressure.
- The effect of the rotation of the Earth – moving air is deflected in a clockwise direction in the northern hemisphere). This is often known as the **Coriolis effect**. In the southern hemisphere moving air is deflected anticlockwise.

In many parts of the world, the distribution of high and low pressure regions is often fixed for much of the year. In turn this means that wind very often blows from a specific direction. In Britain and much of north-western Europe this is from SW to NE.

Coriolis effect (or Coriolis force)

A force caused by the rotation of the Earth which deflects moving air as it moves from high to low pressure.

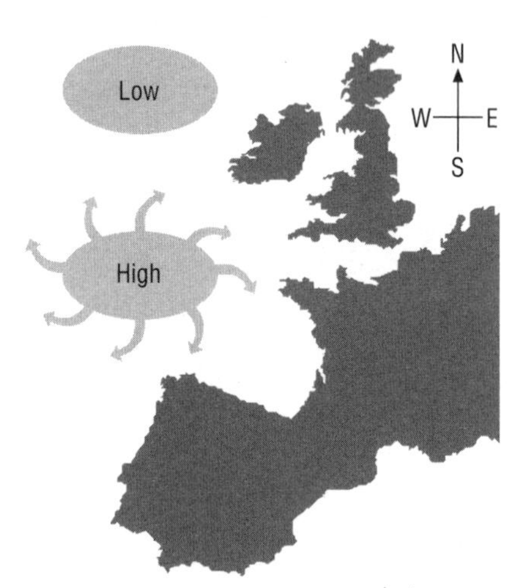

Figure 1 Pressure map of the north Atlantic showing the effect of pressure difference and the Coriolis effect on wind direction

Vertical air movement

The moving air in the atmosphere possesses kinetic energy (the water vapour in the air also possesses chemical potential energy). The original source of all the energy is the Sun. Because energy is always being transferred out of the atmosphere by processes such as friction, there must be a constant input of energy from the Sun.

As seen in spread 2.1.1, most of the energy from the Sun is received by the equatorial and tropical regions of the Earth.

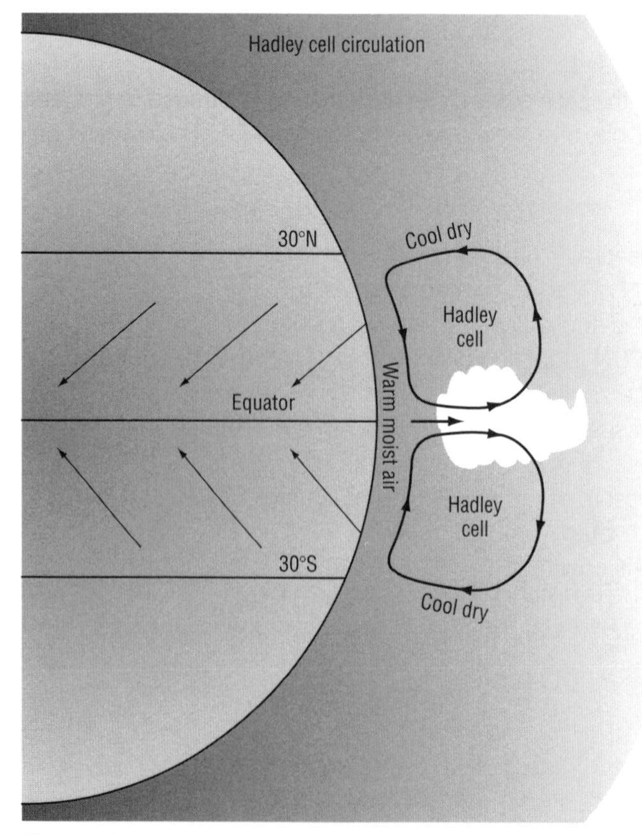

Figure 2 Vertical and horizontal movements of air in a Hadley cell

Examiner tip

If you are asked to predict the wind direction at a given point on the Earth's surface, try to imagine the path of a pocket of air moving from high to low pressure – as it does so it is deflected to the right.

On weather maps showing areas of high and low pressure, the direction of the wind is often indicated by the isobars, which join up areas of equal pressure.

Winds are always named after the direction from which they blow. So a wind blowing from the SE towards the NW is called a south-easterly wind.

This energy is absorbed by the Earth's surface, which then heats up. Air in contact with this hot ground is heated causing it to expand. The molecules become further apart and so the density is reduced. This air is then less dense than the air above it so it begins to rise – the rising of hot air is known as *convection*.

The pocket of rising air cools and eventually it reaches the top of the troposphere – known as the tropopause. At the tropopause, the rising air spreads out horizontally. It returns to ground level by descending at at tropical latitudes (30°N and 30°S). In the process, it warms again. Some of this air then returns at ground level towards the equator.

The returning ground level air from the northern and southern hemispheres meets (or converges) at a region of the Earth's surface called the *inter-tropical convergence zone* (ITCZ). The position of the this region varies over the year and is a major factor in determining the climate of tropical regions.

This cycle of moving air is known as a *Hadley cell*. Similar cells exist between 60° and 90° (a *polar cell*) and, more weakly, between 30° and 60°.

Water vapour and atmospheric circulation

During the circulation described above, changes occur in the amount of water vapour in the air. When air rises and cools, such as at the ITCZ, cooling of the air causes condensation and clouds form.

Descending air, which is warmed as it descends, is associated with evaporation of water vapour and so the air becomes dry and free of clouds. As air returns to the ITCZ, over warm seas or tropical rainforest, it picks up a lot of water vapour by evaporation or transpiration from plant leaves. This means that air arriving at the ITCZ is very moist.

Rising air, causing condensation and rain, is associated with areas of low pressure; descending air, which results in dry air and a lack of clouds, is associated with areas of high pressure.

Applications: hurricanes

The formation of hurricanes and tropical storms is a special case of rising air giving rise to a low pressure region. In this case, the condensation of water vapour that has evaporated from a warm sea releases large amounts of heat, which helps to create a fast moving circulation of air around the centre of the hurricane.

Examiner tip

You may be expected to use the principles in this spread to predict or explain the weather from the information in satellite images or weather maps showing surface air pressure.

Questions

1 What are the two factors that affect the horizontal movement of air across the Earth's surface?

2 What is the ultimate source of the kinetic and potential energy in the atmosphere?

3 **(a)** What is the prevailing wind direction in your geographical region?
(b) Suggest ways in which this wind direction affects the climate and weather you experience.

4 Download a map showing the pressure of the air masses in your geographical region. Use ideas about the two factors that affect the movement of air to predict the direction in which air will move horizontally in your location. Check your prediction by observing the wind direction outside the classroom.

5 Draw a labelled diagram to show the processes connected with water vapour in the circulation of a Hadley cell.

6 Download a current satellite image of a hemisphere of the Earth. Identify the ITCZ and also areas where air is descending (high pressure regions). You may like to compare the positions of these areas with those from an image taken at a different time of year.

Water and covalent bonding

Molecule

A particle made up of two or more atoms bonded together by covalent bonds.

Covalent bond

A type of bond in which atoms share electrons. It is often represented by a single line in structural formulae.

The structure of water

Water is one of the simplest chemical substances in the Universe, consisting of **molecules** in which two atoms of hydrogen are **covalently bonded** to one atom of oxygen. Yet this simple structure has a remarkable set of properties that makes it perhaps the most important substance on Earth for living organisms.

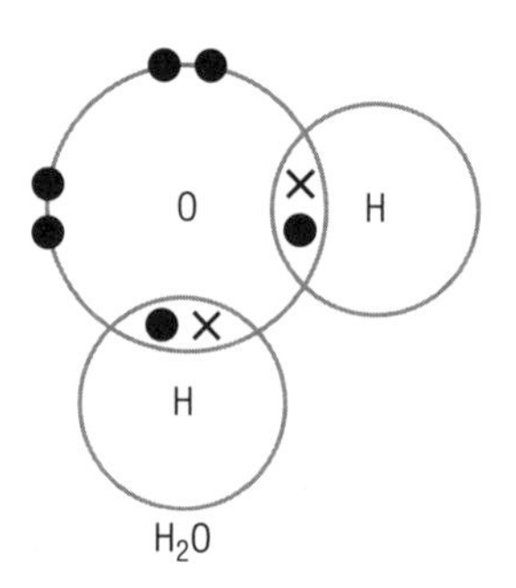

Figure 1 Electronic structure of a water molecule – the dots and crosses show that the electrons being shared come from different atoms

Covalent bonding and electrons

All atoms are made up of a positive nucleus surrounded by shells of negative electrons. It is normally difficult for the nuclei of two atoms to approach each other too closely because the positive charges repel each other. However, in a covalent bond an electron from each of the atoms is shared to form a shared pair of electrons between the two nuclei. The negative charge of this shared pair attracts the two positive nuclei together.

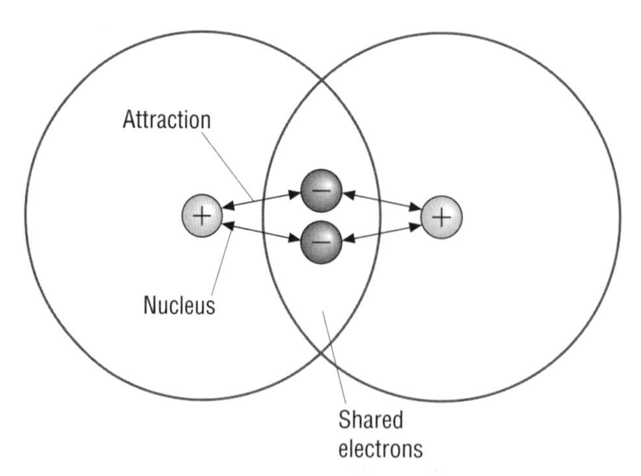

Figure 2 Electrostatic forces in a covalent bond

Electron-pair repulsion theory and shapes of molecules

The shape of molecules can be predicted using the electron-pair repulsion theory. This can be stated as follows:

'Pairs of electrons in the outer shells of atoms repel each other and will be as far apart as possible.'

Examiner tip

When you are trying to work out the shape of a molecule, look for the central atom. This will normally have 8 electrons (4 electron pairs) in its outer shell. So these electron pairs, and any atoms attached to them, will be arranged in a tetrahedral structure.

For almost all the simple molecules that you will come across, there are *four pairs* of electrons in the outer shell of the central atom. This means that the electron pairs repel each other to form a *tetrahedral* shape, with bond angles of about 109°.

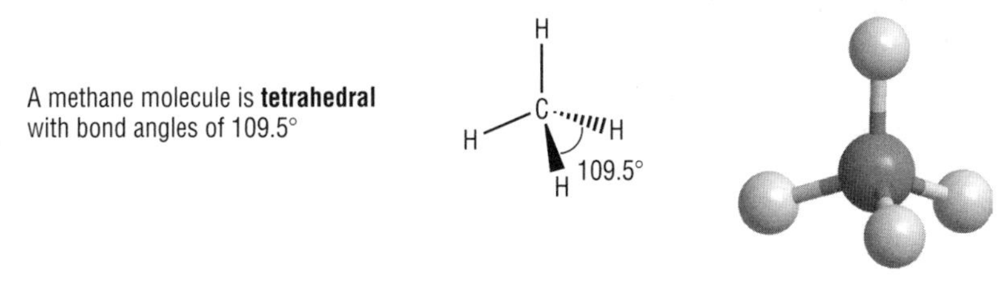

Figure 3 Structure of methane – note how the three-dimensional structure is shown: the 'wedge' bond points forward, out of the paper; the hatched line bond points backwards, into the paper

The effect of non-bonding pairs

Not all molecules that should have a tetrahedral arrangement actually look tetrahedral because many contain atoms with pairs of electrons which are not attached to a second atom – these are known as *non-bonding pairs* (or *lone pairs*). Although the outer electron pairs in these molecules are still arranged in a tetrahedral structure, the atoms appear to be arranged in a different shape. The bond angles between the bonds can still be predicted to be 109°.

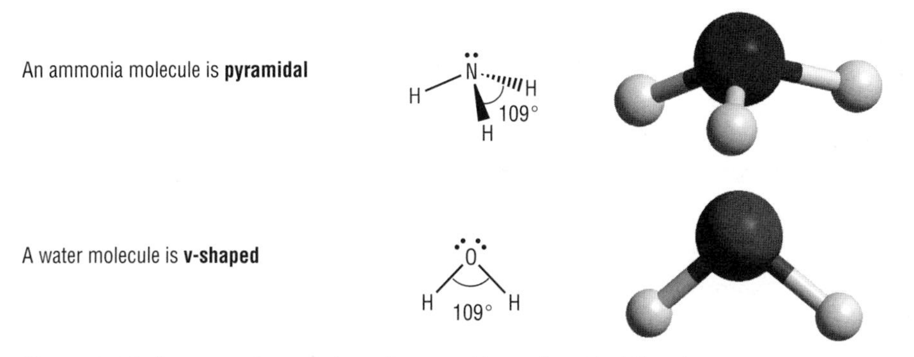

Figure 4 3D diagrams of ammonia and water with predicted bond angles

> **Activity**
>
> Use a set of molecular models to see the shapes of water, ammonia and methane in three dimensions. Some molecular modelling kits include components to represent lone pairs – you will be able to see that in all three cases the arrangement of the electron pairs is tetrahedral.

Note: if you look up the actual bond angles in a water or ammonia molecule you will find they are slightly less than the tetrahedral angle of 109° (107° in ammonia, 104° in water). This is because the lone pairs repel other pairs of electrons slightly more strongly than the bonding pairs of electrons.

Questions

1 The molecular formula of water is H_2O. It shows the number of atoms of each element present in a molecule of water. Write out the molecular formulae of ammonia and methane (the structures of these molecules are shown in Figures 3 and 4).

2 Hydrogen sulfide, H_2S, has a very similar structure to water.
  (a) Draw a dot–cross diagram to show the bonding in hydrogen sulfide.
  (b) Draw a 3D diagram to show the shape of a hydrogen sulfide molecule, indicating the bond angle.

3 Boron trifluoride, BF_3, has an unusual structure in which the boron atom is bonded to three fluorine atoms – there are no non-bonding pairs around the boron atom, so there are only three pairs of electrons around the boron atom.
  (a) Draw a dot–cross diagram of a BF_3 molecule.
  (b) Predict the shape of a BF_3 molecule – draw a diagram to show this shape, indicating the size of the bond angles.

2.1 (5) The polar water molecule and its unusual properties

Electronegativity

Covalent bonding involves electrons being shared. However, electrons are not always shared equally between the two atoms. Some atoms, particularly those that are small or have many protons in their nucleus, attract electrons more strongly than others. We can use **electronegativity** values to predict if electrons will be shared unequally. Table 1 shows some values of electronegativity for common atoms.

Electronegativity

A measure of how strongly an atom attracts electrons in a covalent bond. The higher the electronegativity, the stronger the attraction for electrons.

Atom	C	H	O	N	F	S
Electronegativity	2.5	2.2	3.5	3.0	4.0	2.5

Table 1 Electronegativity in some common atoms

Polar bonds

If electrons are shared unequally then one end of a bond will be slightly negatively charged (written as δ–) and the other will be slightly positive (written as δ+). A bond which has opposite charges at its two ends is described as being *polar*.

Activity

You can investigate whether a liquid is polar or not by finding out if it is affected by electric fields. Rub a plastic rod with a duster and place it close to a stream of the liquid – for example water running out of a burette. If the liquid is polar you will see the stream deflect noticably.

δ– δ+
O—H

Electrons attracted to O atom

Figure 1 A polar O–H bond

The unusual properties of water

Water has several remarkable properties.

- It has an unusually high melting temperature (273 K) and boiling temperature (373 K), considering the small size of the molecule.
- It has an unusually high **specific heat capacity** ($4.2\,J\,K^{-1}\,g^{-1}$). This means that it takes a lot of energy to increase the temperature of a sample of water. You will use this data later on in spread 2.4.1 *Burning fuels*.
- The density of solid water (ice) is less than the density of liquid water at the same temperature – this means that ice floats in water.

Specific heat capacity

The amount of energy required to raise the temperature of 1 g (1 cm^3) of water by 1 °C (1 K).

Density

The mass of a fixed volume of a substance:

$$\text{density} = \frac{\text{mass}}{\text{volume}}$$

Units are $g\,cm^{-3}$ or $kg\,m^{-3}$.

Explaining the unusual properties of water

Most of these properties can be explained because the polar bonds means that water has much stronger intermolecular forces than are normally present between small, simple molecules. Intermolecular forces are the forces between the molecules in a substance and these are always much weaker than the covalent bonds holding the atoms together. It is the intermolecular forces which have to be overcome when the molecules separate – for example when the substance melts or boils. These forces also have to be overcome when the molecules are made to vibrate more energetically when the substance is heated up.

More about intermolecular forces – hydrogen bonding

The relatively strong intermolecular forces between water molecules are known as *hydrogen bonds*. They occur because of three important features of water molecules:

- the H atoms are δ+ and very small
- the O atoms are δ– (because they are electronegative) and also relatively small
- there are two lone pairs of electrons on the oxygen atoms.

These features mean that the hydrogen and oxygen atoms in neighbouring molecules can get very close to each other, strengthening the electrostatic attraction between the δ+ and δ– atoms. A lone pair on the oxygen atom can even be partially shared with hydrogen atoms on neighbouring molecules, making it almost like a covalent bond – this is hydrogen bonding. Notice that the angle between the hydrogen bond and the O–H bond in the water molecule is 180° (Figure 2).

The hydrogen bonds between molecules of water are continually breaking and reforming, which means that the arrangement of water molecules in liquid water remains random. This isn't true in solid water (ice).

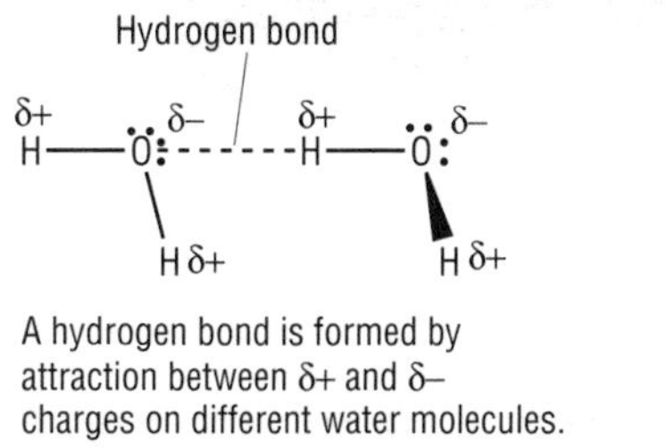

Hydrogen bond

δ+ H—O: δ– - - - - H δ+ —O: δ–

H δ+ H δ+

A hydrogen bond is formed by attraction between δ+ and δ– charges on different water molecules.

Figure 2 Hydrogen bonding between two water molecules

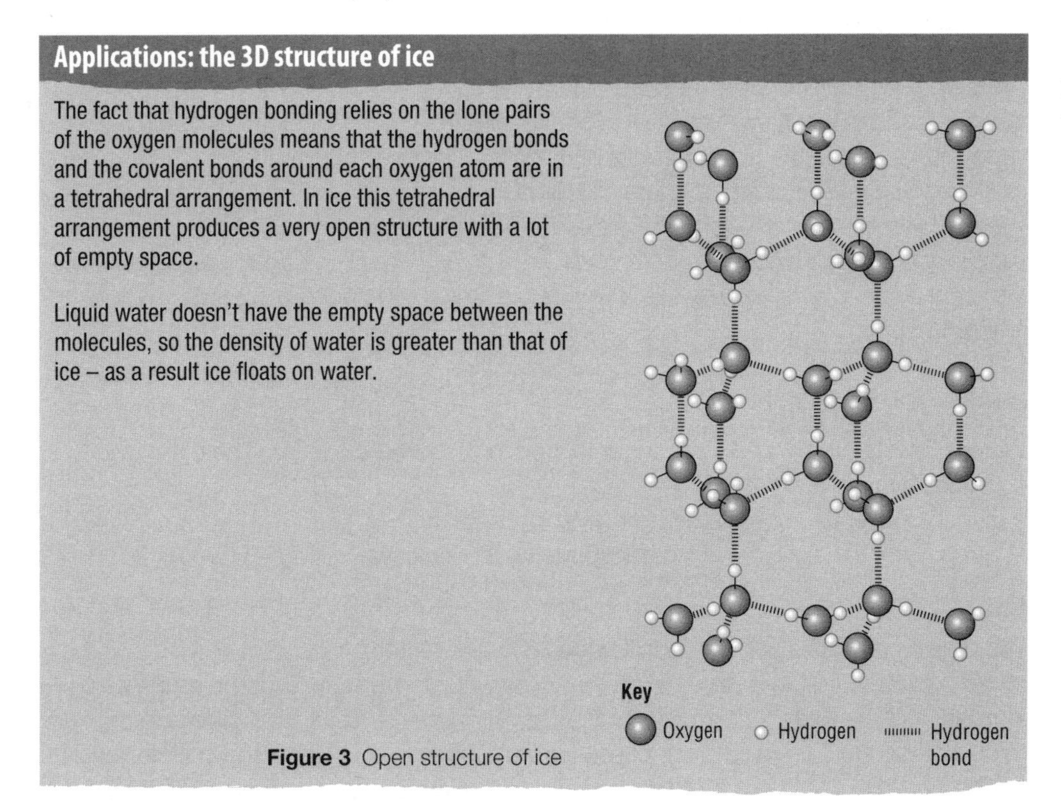

Applications: the 3D structure of ice

The fact that hydrogen bonding relies on the lone pairs of the oxygen molecules means that the hydrogen bonds and the covalent bonds around each oxygen atom are in a tetrahedral arrangement. In ice this tetrahedral arrangement produces a very open structure with a lot of empty space.

Liquid water doesn't have the empty space between the molecules, so the density of water is greater than that of ice – as a result ice floats on water.

Figure 3 Open structure of ice

Condensation and evaporation

In spread 2.1.3 you saw how evaporation requires energy, whereas condensation (where water vapour becomes liquid water) releases energy. The energy needed for evaporation breaks the hydrogen bonds between the molecules, whereas hydrogen bonds are formed when condensation occurs – and forming bonds releases energy.

Questions

1 List three unusual properties of liquid or solid water.

2 Use electronegativity values to predict if the H–Cl bond (in hydrogen chloride molecules) is polar.

3 Methane (CH_4) is non-polar, whereas ammonia (NH_3) is polar.

  (a) Draw diagrams to show the structure of these two molecules and explain this difference in properties.

  (b) Which substance would you expect to have the higher melting and boiling temperatures? Check your answer in a data book.

4 Explain in terms of the bonds in the molecules why:

  (a) water requires a large amount of energy to boil it

  (b) a lot of energy is required to heat water up to 100°C.

Ocean circulation

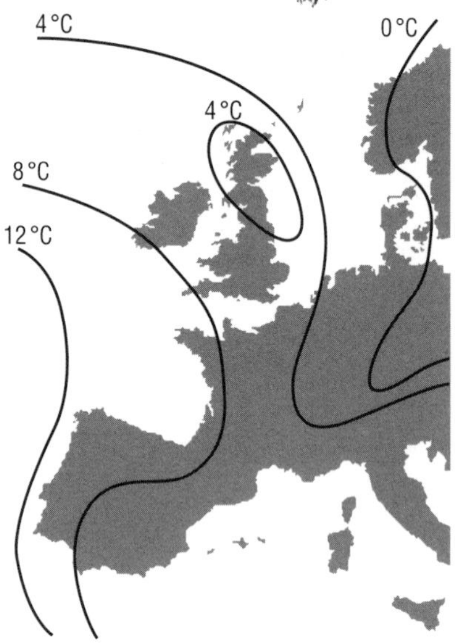

Figure 1 January isotherms for the north Atlantic

Activity

Find data for the June and January temperatures in the following cities, which are on roughly the same latitude:

- Cardiff
- London
- Berlin
- Winnipeg.

Then use an atlas to find their geographical locations and use ideas about the moderating influence of the oceans to explain the differences you find.

Examiner tip

You will need to be able to remember the pattern of prevailing winds you studied earlier in this module in order to explain the origin of the sub-tropical gyres.

Similarly, you will need to be able to recall the pattern of deflection caused by the Coriolis effect in order to predict the direction of rotation of these gyres.

Oceans and climate

The Earth's oceans have a huge impact on the climate of many parts of the world. The presence of large volumes of water helps to maintain a relatively constant mild temperature for much of the year in coastal regions, including the UK. On top of this, the global circulation of warm and cold water is a major factor in determining the climate experienced over large areas of the Earth's surface. When these patterns are disrupted, unusual, and sometimes devastating, climatic shifts occur.

The moderating influence of the oceans

Water has an unusually high specific heat capacity. This means that it heats up relatively slowly compared to the soil and rocks of the land. Relatively high amounts of energy are required to increase the temperature of water significantly. Similarly, water is slow to cool – it needs to lose a lot of energy before its temperature drops significantly. So the UK is surrounded by seas whose temperature does not vary by more than a few degrees over the course of the year, keeping the UK milder in winter and cooler in summer than mainland Europe.

Ocean circulation

Water is constantly circulating in currents around the world's oceans. There are two components to this circulation:

- surface currents – which transport water horizontally
- deep water circulation – which transports water vertically.

Surface currents

The direction of these currents is influenced by three factors:

- prevailing winds – these drive the movement of water in the same direction as the winds
- rotation of the Earth – as in the atmospheric circulation system, the Coriolis force, caused by the Earth's rotation, causes ocean currents to be deflected in a clockwise direction in the northern hemisphere and anticlockwise in the southern hemisphere
- the presence of land masses – currents which are driven by the winds against land masses cause water to pile up against the edge of the land, and then to gradually flow out parallel to the coast.

The overall effect of these features is to produce several strong spiral currents. These are particularly noticeable close to the equator where they are known as *sub-tropical gyres*.

Applications: warm and cold currents and El Nino

You should be able to see from the pattern of the currents in the sub-tropical gyres that some of the currents (those flowing from high latitudes towards to equator) are likely to be cold – others, flowing away from the equator, will be warm.

In particular, you will notice that there is a cold current flowing up the west coast of South America., In recent years, however, this current has often disappeared to be replaced by a warm current flowing south. The whole south Pacific gyre seems to go into reverse, reversing the distribution of high and low pressure systems and altering the weather in the Pacific, Australia and the Americas in a dramatic way. This is known as an 'El Nino' event. The current may oscillate back and forth in this way several times a decade.

It is now thought that changes in water temperature in the North Atlantic may also cause the changes in the high and low pressure systems there. This 'north Atlantic oscillation' does not have as dramatic effects as El Nino but can help us to understand why, for example, some winters in the UK are much wetter and milder than others.

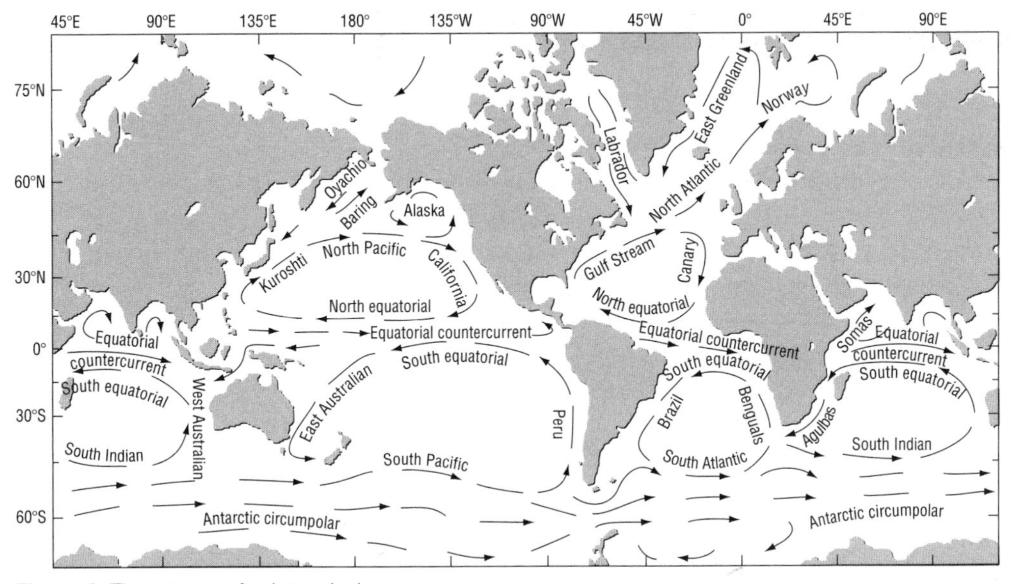

Figure 2 The pattern of sub-tropical gyres

Currents in the north Atlantic

Of particular significance to the UK are the series of currents which occur as a result of the sub-tropical gyres in the Atlantic Ocean.

The prevailing south-easterly winds blow water from the coast of west Africa across to the Gulf of Mexico. Some of this becomes part of the sub-tropical gyre shown in Figure 2, but much of it is forced into the Gulf of Mexico which is very warm and shallow. The water heats up – sometimes generating powerful hurricances – and eventually this warm water is forced out along the eastern coast of the United States where it is called the Gulf Stream.

It is deflected clockwise and begins to cross the Atlantic where it is deflected clockwise towards Europe as the north Atlantic drift. Although it begins to break up and split into separate currents as it passes the British Isles, one strong branch passes to the north-west of the UK and Norway as the Norwegian current. The reason for the strength of the current so far north is that just south of Greenland the water begins to sink. This drives the series of currents in a kind of conveyor belt. The sinking is discussed in the next spread.

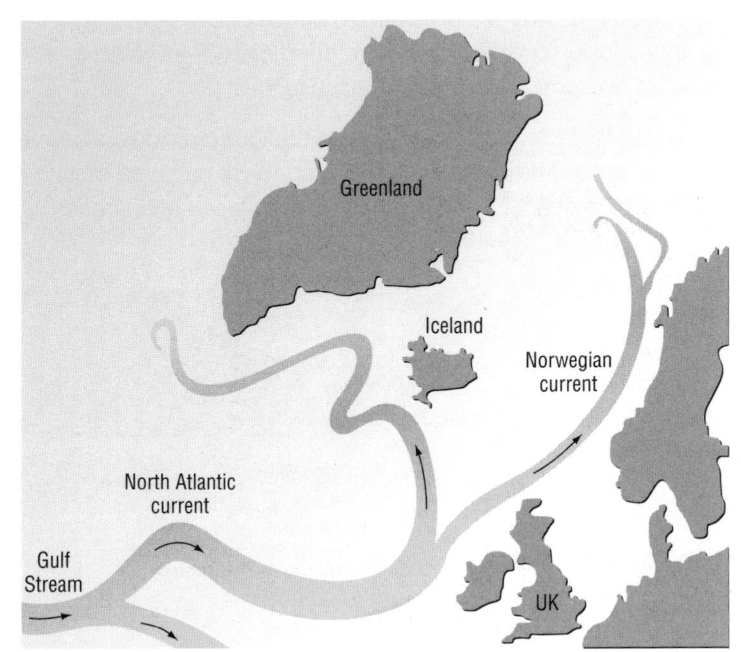

Figure 3 The Gulf Stream and associated currents

Activity

Find out if the Pacific Ocean is currently in an El Nino event (the US monitors this carefully – search for 'El Nino + NOAA'). The UK Met Office issues forecasts for coming seasons using data for the north Atlantic oscillation – find the latest predictions on their website.

Questions

1 Which property of water means that it cools down and warms up much more slowly than many other substances?
2 List three factors which affect the surface movement of water.
3 In what direction does water circulate in sub-tropical gyres in the southern hemisphere? What happens to this direction in the Pacific Ocean during an 'El Nino' event?
4 Warm water currents, such as the north Atlantic drift, make the climate of some parts of the world warmer. Suggest and explain one other effect that they may have on the climate of a region.

2.1  The thermohaline circulation

Vertical circulation of water

You saw in spread 2.1.6 how surface ocean currents arise and their importance in determining climate. There is a second, equally important, circulation system which is driven by vertical movement of water – in particular the sinking of water at two regions of the Earth – this is known as the **thermohaline circulation**.

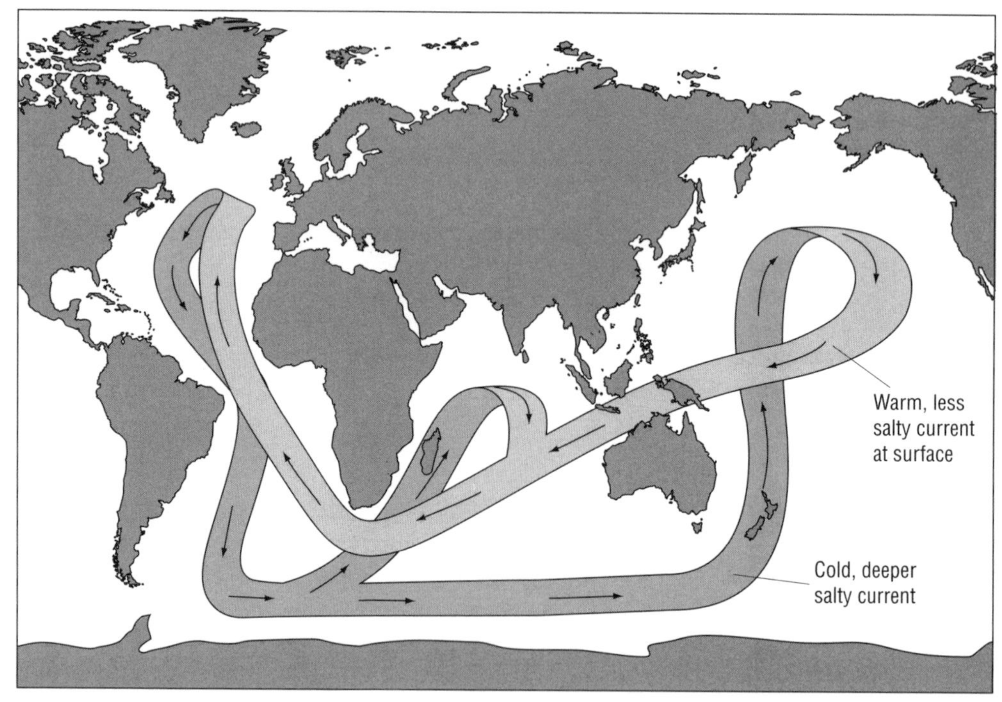

Figure 1 Thermohaline circulation

Density of sea water

The density of a substance depends on two factors – how closely packed the particles are and the mass of the particles.

When a liquid is cooled it usually becomes more dense because the particles, having less energy, become closer together – this is rather more complicated in the case of water as the effect of hydrogen bonding can oppose this.

Sea water is a solution that contains salts such as sodium chloride and potassium bromide dissolved in water. The ions which make up these salts are much heavier than molecules of water so the effect of adding salts to the water is to increase its density.

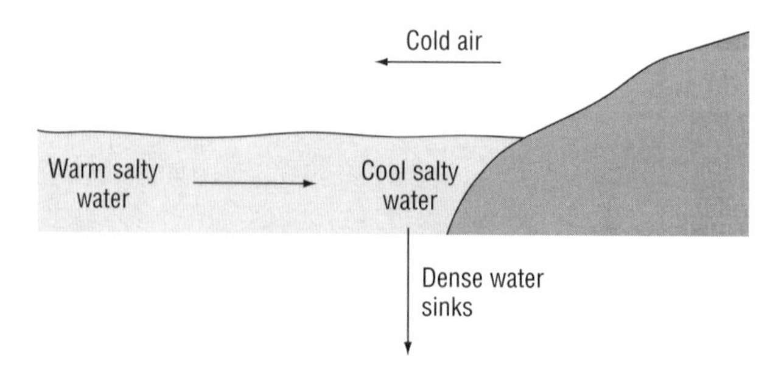

Figure 2 The sinking process in the Norwegian Sea

Thermohaline circulation

Sinking or rising of water caused by differences in temperature and concentration of dissolved salts (particularly halides).

Worked example

1 m^3 of sea water contains, on average, about 35 kg of dissolved salts. It also contains about 1000 kg of water.
The density of sea water is the mass of 1 m^3 of sea water – sea water therefore has a density of 1035 $kg\,m^{-3}$.
The Gulf Stream flows at a rate of 30 000 000 m^3 every second.
We can use the fact that mass = density × volume to calculate the mass of water flowing every second in the Gulf Stream – it is $1035 \times 30\,000\,000 = 3.11 \times 10^{10}$ kg.

Sinking in the Norwegian Sea

As the Gulf Stream flows across the north Atlantic, a lot of water vapour will evaporate from the surface. The dissolved salts do not evaporate, so gradually the concentration of salt in the water increases – this increases the density. As it reaches the Norwegian Sea (a region between Greenland and Norway) and becomes the Norwegian current, the water is cooled by the cold winds blowing over the sea. This cooling makes the water in the Norwegian current denser than the surrounding water and it sinks.

There is some concern that if meltwater (which contains no salt) from melting glaciers dilutes the salty water, the density may not be high enough for the water in the Norwegian current to sink. Because it is the sinking process which helps to drive the Gulf Stream across the Atlantic some scientists fear that the Gulf Stream, and the other currents which derive from it, may weaken or even stop altogether in future, particularly if global warming causes rapid melting of the glaciers.

Sinking in the Weddell Sea

A different process causes sinking of water in the southern hemisphere. This occurs off the coast of Antarctica in a vast bay called the Weddell Sea. Ice forms rapidly here at the beginning of winter. When ice forms, however, it is only the water molecules which freeze to form the ice structure – the particles of salt are left behind in the seawater. As a result, the concentration of salt increases and the density becomes so great that the water begins to sink. Again, if global warming affects the formation of ice, the sinking in this region may be affected.

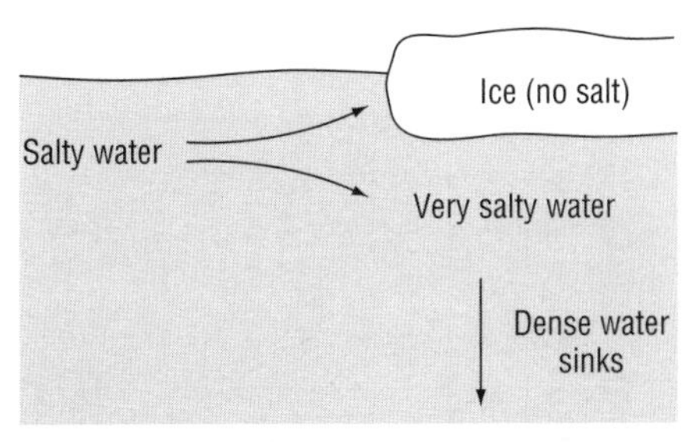

Figure 3 Sinking process in the Weddell Sea

Upwelling of water (the rising of water from the ocean floor)

This is much harder to detect than the sinking of water but it is thought to occur mostly in the north Pacific. One of the effects of upwelling is that it can bring nutrients from the bottom of the ocean to the surface and so may help to increase the productivity of marine ecosystems. Scientists think that it may take a hundred years or more for water to resurface after sinking. The sinking water takes with it a great deal of dissolved carbon dioxide. So carbon dioxide from the time of the Industrial Revolution may now be reaching the surface of the ocean in upwelling water where it may be released back into the atmosphere, adding to global warming.

Questions

1. Explain why water from the Gulf Stream is so dense when it reaches the Norwegian Sea. What happens in the Norwegian Sea to make it dense enough to cool?
2. Which is likely to be saltier – sea ice or the unfrozen sea water beneath it? Explain your answer.
3. The Dead Sea, an inland lake, contains 350 kg of dissolved salts in every 1 m^3.
   (a) Assuming that 1 m^3 of water from the Dead Sea contains 1000 kg of water, calculate the density of water in the Dead Sea. Do you think your assumption is justified?
   (b) What volume of water from the Dead Sea will have a mass of 1 kg? Use the figure for density from part **(a)**
   [Hint: density = $\frac{\text{mass}}{\text{volume}}$ so you may need to rearrange this]
4. **(a)** Why do you think water from the ocean floor is likely to be rich in nutrients?
   (b) It been suggested that we can encourage marine ecosystems to be more productive by artificially raising water from the deep ocean. Is this likely to be a good idea? Consider, for example, the possible effects of global warming.

Evidence for climate change

Greenhouse gases and global warming

You may see articles in newspapers or on the Internet which claim that there is uncertainty about whether the increased emissions of greenhouse gases (particularly carbon dioxide) are causing global warming or not. In reality, most climate scientists believe that there is overwhelming evidence to suggest there is a connection between the two. The facts are these:

- There has been an unusually rapid rise in global temperatures since the middle of the twentieth century.
- This has occurred at the same time as carbon dioxide concentrations in the atmosphere have increased rapidly.
- Scientists know that there is a mechanism by which increased carbon dioxide concentration will cause an increase in atmospheric temperature (see spread 2.2.9 *Greenhouse gases*). They have developed **models** which allow them to predict how changes in carbon dioxide concentration will affect temperature – these models closely match what has been observed happening in the atmosphere.

Scientific model

An idea that allows us to create explanations of how we think some part of the world works. A model may be based on mathematical patterns and can be used to make precise predictions about the effect of changes in conditions.

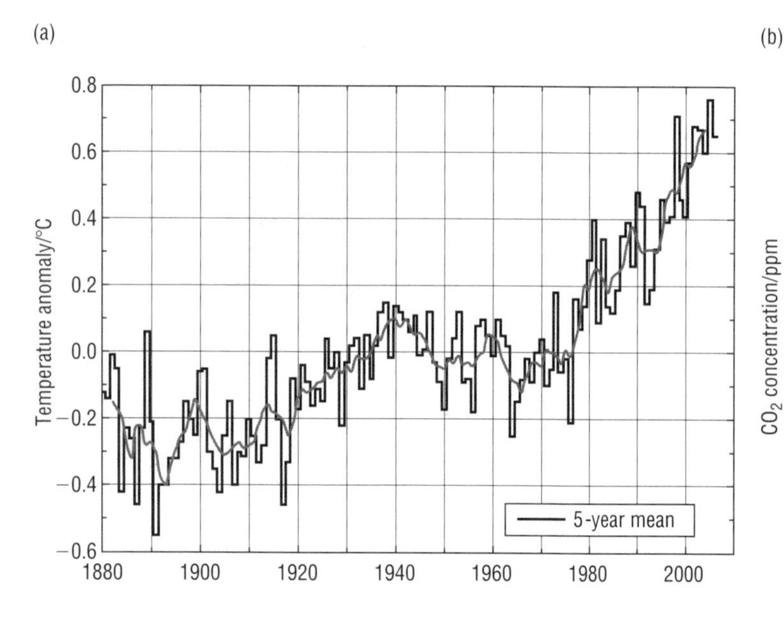

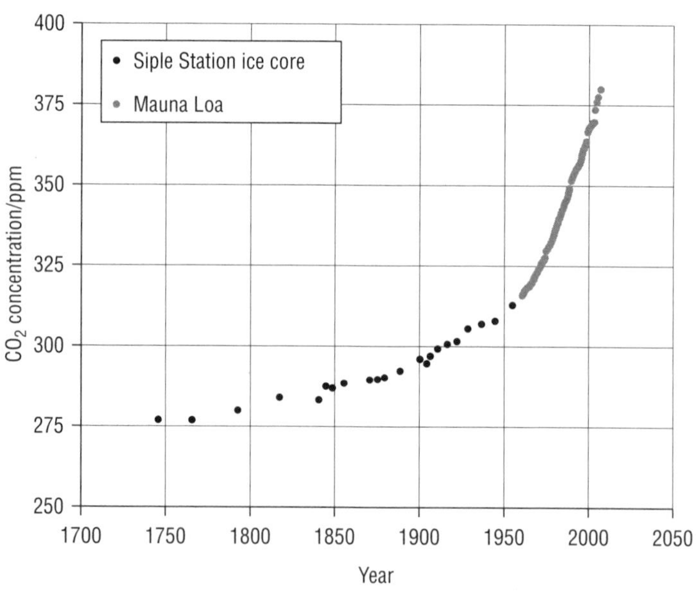

Figure 1 Climatic changes: **a** global temperature; **b** global CO_2 concentration

The problem is that direct measurement of temperature dates back a few hundred years – and often only for quite localised regions of the Earth's surface, while direct carbon dioxide measurements have only been made in the past century.

Extending the data back

Although direct temperature measurements only cover a few centuries, it is possible to use indirect (or proxy) evidence to make an estimate of the temperature in historical times.

Tree-ring analysis

This is based on the well-known idea that the amount of growth shown by a tree each year can be easily observed in the form of a 'ring'. The thickness of a ring gives an idea of how suitable the weather was for tree growth that year. Tree rings can be dated accurately, sometimes by using **radiocarbon dating**.

Radiocarbon dating

A way of estimating the age of material that was once alive. It relies on the fact that the radioactivity of the ^{14}C isotope falls in a regular way after the death of an organism.

Pollen analysis

Pollen grains are often preserved in peat bogs, lake sediments etc. and can often be dated fairly accurately – pollen grains at the bottom of sediments will be the oldest. Analysing the pollen gives information about the type of vegetation present at that time, which in turn will suggest what the climate was like. Spread 2.2.3 *Clearly a problem* shows another use of pollen-grain analysis.

The hockey stick graph

Use of these techniques has produced some striking evidence to suggest that, looking at global temperatures, the Earth has warmed much more dramatically recently than at any time in the past thousand years – shown by the sharp curve upwards in the last 50 years, looking like the end of a hockey stick.

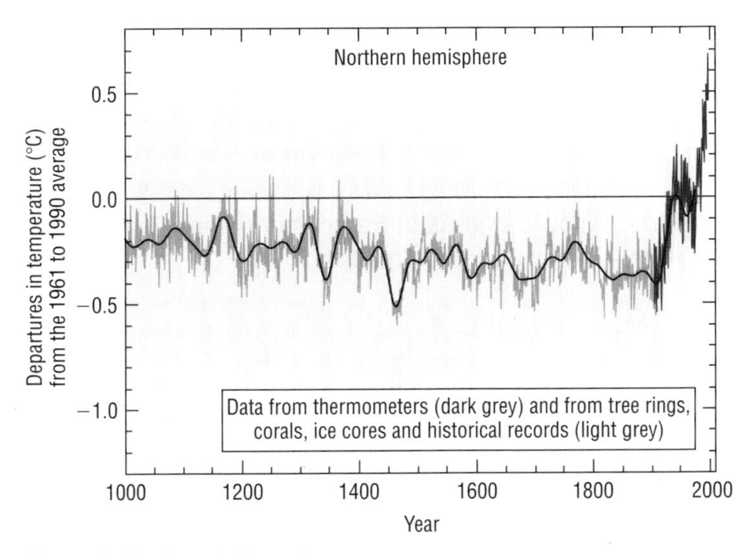

Figure 2 'Hockey stick' graph

Further back into the past: ice ages and interglacials

One of the most frequently heard arguments used against the model of man-made global warming is that the climate in the past 200 000 years has been extremely variable, with ice ages being followed rapidly by warm interglacial periods. Modern techniques now make it possible to show exactly how the temperature has varied over that period.

Ice-core sampling

The thick layers of ice in the Arctic and Antarctic act as a remarkable record of thousands of years of climate. Each year's snow contains water frozen out of the atmosphere. The amount of an unusual **isotope** of oxygen, ^{18}O, present in the frozen water allows scientists to make an estimate of the temperature of the atmosphere – a warm atmosphere will contain more ^{18}O in the form of water vapour.

A long core of ice, as long as a kilometre, is removed from the ice and analysed to measure the different amounts of ^{18}O at different depths. Like tree rings, the layers of ice can be dated accurately.

It is also possible to analyse samples of the atmosphere from bubbles of gas trapped in the ice cores. This can provide information about the amounts of greenhouse gases present in the atmosphere at different times. You will carry out some analysis of these data in spread 2.1.10 *Analysing climate data*.

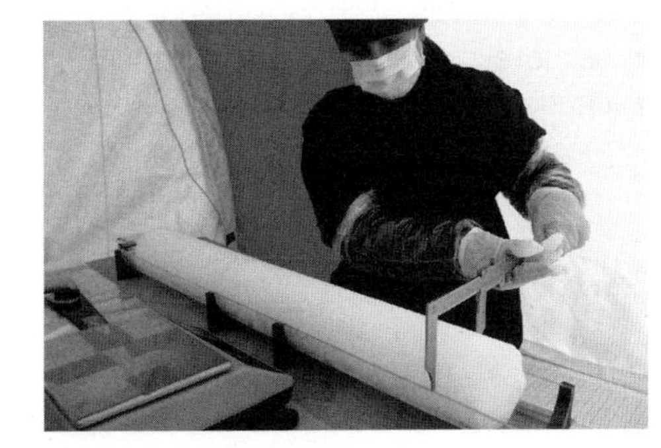

Figure 3 Analysing an ice core

How science works

In this spread you will examine and evaluate how scientists have used a range of techniques to provide evidence for climate change (HSW 5c).

Activity: debates over the evidence

The climate change debate is one of the most interesting to look at in order to see how data and evidence are evaluated and how conflicting evidence is handled by the scientific community.

Searching the Internet or reading newspaper databases will produce many articles that attempt to cast doubt on some of the evidence used to support the generally accepted model of man-made global warming. Look at some of these sites and decide whether the assumptions and reasoning are valid. If so, does this mean that we should reject the current model of global warming?

Isotopes

Different forms of the same type of atom that have the same atomic number but different mass numbers.

Questions

1 (a) Explain why tree rings are likely to be thicker in a warmer year.
(b) What other factors might also affect the thickness of tree rings, apart from temperature?

2 List any other evidence which scientists and archaeologists could use to deduce whether the climate in the past was warmer or colder than it is today.

3 What arguments are used by people who disagree with the idea that global warming is happening? How valid do you think these arguments are?

Models of climate change

How science works

In this spread you will look at how mathematical models can be used to predict a range of outcomes of global warming (HSW 7a) and how these models are validated (HSW 7b). You will also consider the ways in which society should respond to the predictions of these models (HSW 7c).

Greenhouse effect

The warming of the atmosphere caused by the ability of some gases to trap infrared radiation emitted by the Earth.

Greenhouse gas

A gas which can absorb and trap infrared radiation emitted by the Earth. Examples include carbon dioxide, methane, water vapour and chlorofluorocarbons.

Applications: feedback loops

You came across positive and negative feedback loops in spread 1.2.8.

Feedback processes which may affect temperature levels include:

- melting ice exposing rock which absorbs sunlight energy more effectively
- more evaporation from oceans causing increased cloud cover in the atmosphere
- higher temperatures and greater carbon dioxide levels increasing plant productivity
- higher temperatures increasing the solubility of carbon dioxide in the oceans
- higher temperatures causing more melting of permafrost and the release of methane trapped in the frozen soil.

You should be able to suggest whether these processes will be negative or positive feedback loops.

Constructing models

The models and simulations which climate scientists use are incredibly complex. They must take into account not only the effect of rising **greenhouse gas** levels (methane, as well as carbon dioxide) but also a range of feedback loops that will come into play as temperatures rise. Positive feedback loops could produce a 'runaway' **greenhouse effect**, producing catastrophic global warming.

Because there are so many different models with slightly different assumptions, some scientists have been reluctant to accept the predictions. One thing that can be done is to run the models into the past to see if, for example, they correctly predict the observed warming in the second half of the twentieth century. Most of the models do so with remarkable accuracy.

The predictions

Running the models through until 2100 (Figure 1) suggests a range of possible outcomes, depending on how successfully we limit the increase in carbon dioxide production and on the exact details of the models.

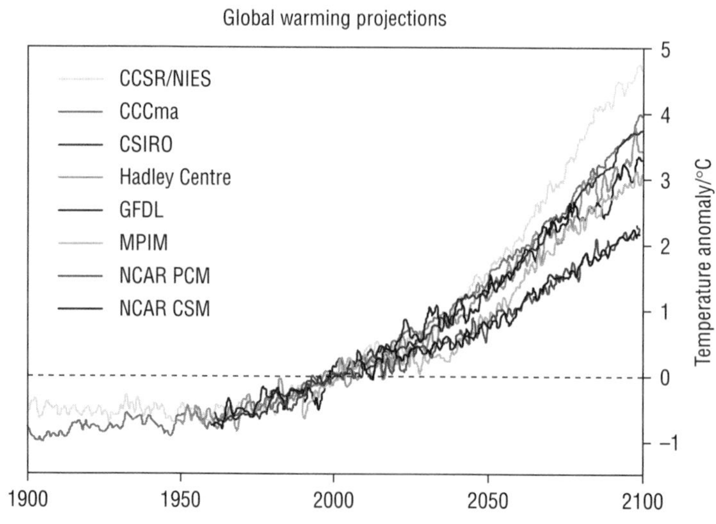

Figure 1 Comparison of the predictions

Another unknown factor is whether the increase in temperature will be even across the globe. Most models suggest that this will not be the case. Notice the exceptionally large increases in the Arctic and the Amazon basin in the model shown in Figure 2. These may create irreversible changes, such as the complete disappearance of the Amazon rainforest and the Greenland ice cap.

Environmental and social effects

The likely effects of climate change have been widely publicised and include:

- sea level rise causing flooding of low-lying areas and habitat destruction
- changes in rainfall patterns causing catastrophic droughts in some areas
- mass extinction of plants and animals with major disruption to habitats
- many parts of the world becoming unusable for agriculture, causing famines
- social unrest caused by the large-scale movement of refugees from areas affected by drought, famine or flooding.

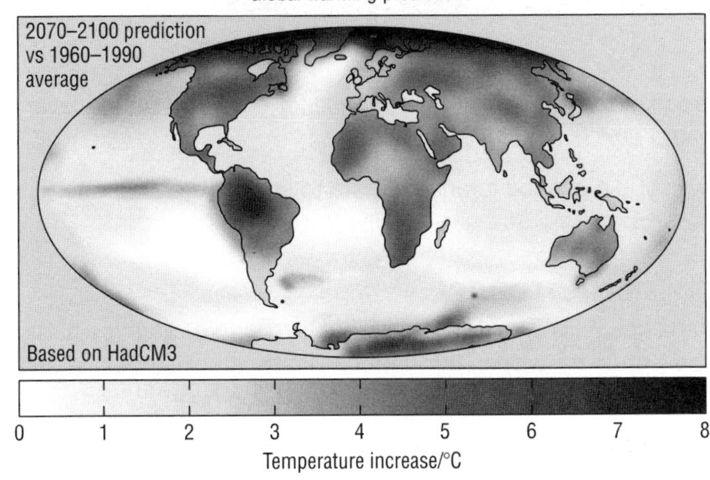

Figure 2 Variation in temperature increase across in different regions of the Earth

Is climate change unavoidable?

It is now clear that, even with current levels of carbon dioxide emissions into the atmosphere, a rise of 1–2 °C is certain to occur over the next 20–50 years. Whether greater increases will occur depend on to what extent we can limit or even reduce the emissions of greenhouse gases. This will be made even more difficult because countries such as India and China are rapidly becoming highly industrialised societies, dramatically increasing their energy requirements.

Strategies for limiting climate change

These could include:

- Using alternative energy sources – these could include renewable energy sources such as wind, wave and solar. It could also mean using biofuels that are derived from plants which absorb carbon dioxide as they grow. Or it could mean an increased use of nuclear power, making use of nuclear fission reactions. You will study this in Module 4.
- Carbon capture – this involves finding a way to trap the carbon dioxide released from burning fossil fuels in some way and then to store it safely so that it is not released into the atmosphere.
- Energy efficiency – technology exists to enable a reduction in the energy required by industry and society, by making more use of recycling, for example, or by designing more efficient machines and processes.
- Political and economic agreements – many people believe that the strategies above are unlikely to be implemented unless there are political or economic agreements forcing societies or individuals to reduce their energy consumption.

Examiner tip

You may be asked to discuss and evaluate these strategies – all of them have potential problems which you should be able to identify by research.

Questions

1 Sea levels will rise if global warming continues. Explain why this will happen. [Hint: there are two different effects involved!]

2 The growing of biofuels to produce a 'carbon-neutral' energy source has increased in recent years. However, some people oppose this. Suggest arguments which might be used against the increased growing of crops to use as biofuels.

3 One suggestion for carbon capture is to liquefy the carbon dioxide and to store it at the bottom of the ocean. Is this suitable as a long-term store for carbon dioxide? [Hint: use ideas from spread 2.1.7 *The thermohaline circulation*].

2.1 (10) Analysing climate data

How science works

In this spread you will interpret data to find patterns and possible causes of climatic change in the past (HSW 5b).

In previous spreads you have looked at evidence for climate change – over the course of this century and also extending back thousands of years.

This activity gives you a chance to analyse data to find if there are patterns in the way climate changes.

1 In and out of the ice ages

In the nineteenth century, scientists realised that the Earth's climate – particularly that of northern Europe – must have been through many dramatic changes. In particular, many geological findings – for example huge boulders that had been carried vast distances from their point of origin – suggested that Britain had been periodically covered with extensive sheets of ice, and so the Earth's climate must have been much colder at that time. We call these periods of time *ice ages*.

Recently, important evidence about the Earth's temperature history has come from the analysis of water molecules in Antarctic ice (see spread 2.1.8) – the data are shown in Table 1. Scientists have made estimates of the difference between the average temperature of the Earth's atmosphere in the past and now. A negative difference means that the atmosphere was much colder and this shows clearly when ice ages occurred. At the same time, analysing bubbles of air in the ice cores (see spread 2.1.8) allows scientists to measure the actual concentrations of carbon dioxide and other gases in the atmosphere at that time.

Age of sample/ years	Temperature difference/°C	CO_2 concentration/ ppm
0	0.0	276
5	−0.1	258
10	−0.8	255
15	−7.7	203
20	−9.0	196
25	−8.6	200
30	−7.7	223
35	−7.2	211
40	−6.0	190
45	−6.9	208
50	−6.6	202
55	−5.7	215
60	−7.8	198
65	−7.6	205
70	−6.5	244
75	−5.9	226
80	−3.4	223

Age of sample/ years	Temperature difference/°C	CO_2 concentration/ ppm
85	−5	216
90	−4.9	228
95	−4.9	232
100	−2.8	228
105	−4.2	240
110	−6.6	237
115	−4.9	276
120	−1.4	277
125	0.1	272
130	1.1	275
135	1.6	267
140	−4.6	225
145	−8.7	191
150	−8.5	199
155	−8.8	193
160	−8.5	196

Table 1

What to do

Use a spreadsheet to analyse the data.

1 Plot a graph of temperature difference against age of sample. Identify the ice ages on your graph. How quickly did the atmospheric temperature change at the beginning and end of ice ages – calculate the rate of change in °C per century.

2 Plot a similar graph of carbon dioxide concentration against age of sample. What do you notice about the patterns in the two graphs you have plotted? What might this suggest about the relationship between temperature and carbon dioxide concentration?

3 To test this, try plotting a scatterplot of temperature difference against carbon dioxide concentration. Does this graph suggest a link between carbon dioxide concentration and temperature?

4 Do your results prove that changes in carbon dioxide concentration cause changes in temperature? What other explanation could there be for your results?

5 Carry out some research to find out what scientists think were possible causes of the ice ages.

2 Predicting droughts

In spread 2.1.1 you learnt that the climate of many parts of the world might be expected to be fairly predictable. For example, the Sahara desert is dry all year round because the air pressure is always high in this part of the world. Many countries in the rest of Africa have distinct wet and dry seasons, depending on whether the air pressure is high or low over that country. For example Zimbabwe, at around 20°S, is normally dry during the winter months (May–August) and receives most of its rain in the summer (November–March) when the ITCZ (see spread 2.1.3) is over the country. However, the movement of the ITCZ is sometimes unpredictable – in some years it may not extend as far south as Zimbabwe and in these years rainfall can be low with serious consequences for farming. The data showing the average rainfall over the country in Table 2 were collected from 1930 to 1993.

Year	Rainfall/cm	Year	Rainfall/cm	Year	Rainfall/cm	Year	Rainfall/cm
1930	568	1946	365	1962	788	1978	569
1931	725	1947	765	1963	467	1979	640
1932	591	1948	535	1964	509	1980	861
1933	565	1949	519	1965	571	1981	440
1934	668	1950	517	1966	677	1982	403
1935	585	1951	784	1967	405	1983	464
1936	656	1952	909	1968	716	1984	746
1937	553	1953	582	1969	539	1985	695
1938	977	1954	1013	1970	577	1986	422
1939	704	1955	763	1971	806	1987	744
1940	627	1956	692	1972	371	1988	605
1941	501	1957	796	1973	1004	1989	625
1942	788	1958	656	1974	820	1990	502
1943	731	1959	483	1975	737	1991	335
1944	601	1960	728	1976	748	1992	380
1945	735	1961	612	1977	981	1993	640

Table 2

What to do

1 Use a spreadsheet to plot a graph of rainfall against year. Can you observe any patterns?

2 The problem with rainfall and temperature data is that the wide variations from year to year may mask any underlying pattern or trend. One solution is to smooth out the variation by taking the mean of several years' data. Adapt your spreadsheet so that it calculates the five-year average of the data on either side – for example, for '1932' calculate the average of the years 1930–1934. If you include any more than five years in the average, any underlying patterns may also be smoothed out. Plot a graph of the five-year average against year. Can you see any patterns now?

3 Droughts (and floods caused by excessive rainfall) will cause problems for farmers and economic difficulties that could lead to food shortages. But other factors may cause famines as well. Find out something about the recent history of Zimbabwe. Do you think the variations in climate are a key factor in the food shortages which Zimbabwe has experienced?

2.1 Module 1 – Summary

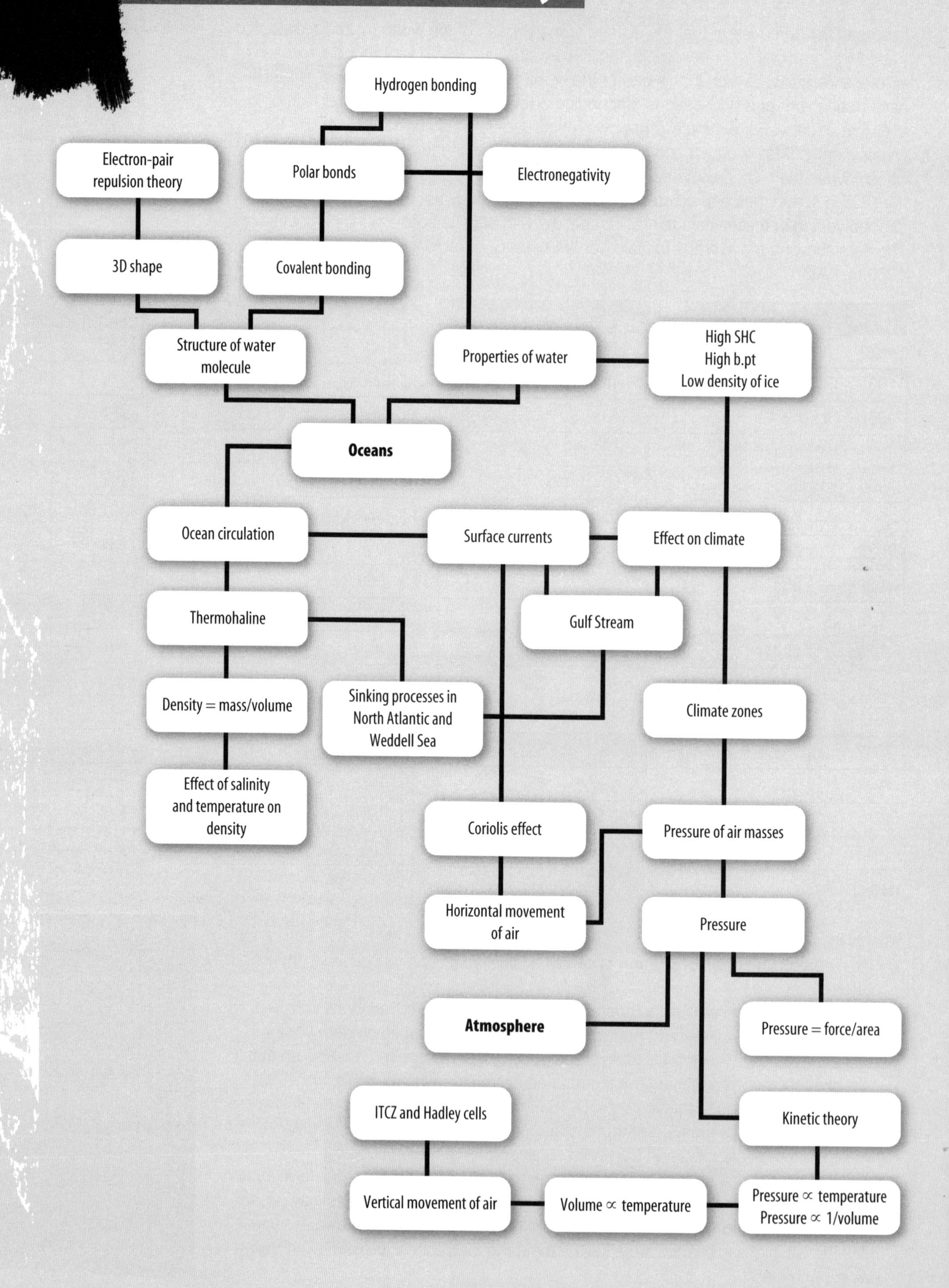

Practice questions

Low demand questions

These are the sort of questions that test your knowledge and understanding at E and E/U level.

1 (a) Use ideas about the movement of molecules to describe how a gas exerts pressure on the walls of a container.

(b) Copy and complete Figure 1 to show the patterns of high and low pressure at different latitudes of the Earth.

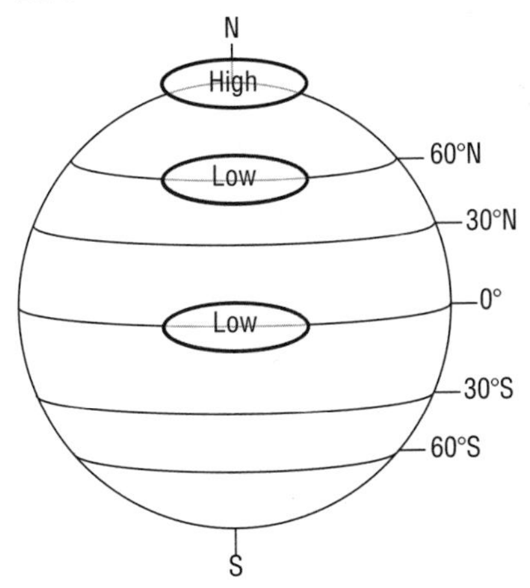

Figure 1 Incomplete diagram showing high and low pressure zones

2 Water is a small molecule with several unusual properties that can be explained by its structure.

(a) State two unusual properties of water.

(b) The structure of water is shown in Figure 2.

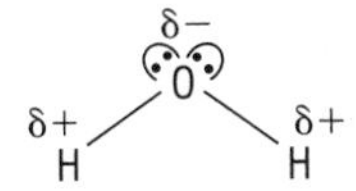

Figure 2 Structural formula of water

Explain what is meant by:

(i) the pair of dots above the O atom

(ii) the single line between the O and the H atoms

(iii) the label δ+ on the H atoms.

Medium demand questions

These are the sort of questions that test your knowledge and understanding at C/D level.

3 Figure 3 shows areas of high and low pressure in the northern hemisphere, as they might be indicated on a weather map.

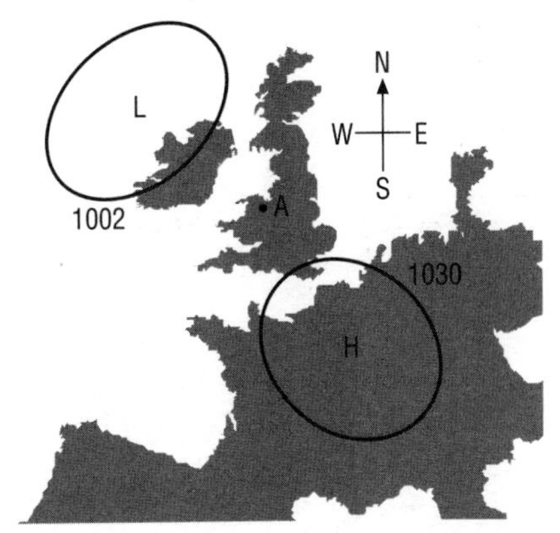

Figure 3 Simplified weather map showing high and low pressure

(a) Predict the likely wind direction at point A.

(b) Explain your answer by discussing the factors affecting wind direction.

4 (a) Identify one region of the northern hemisphere in which sea water is observed to be sinking.

(b) Describe why sinking occurs in this region.

High demand questions

These are the sort of questions that test your knowledge and understanding at A/B level.

5 Concern about the effect that increasing carbon dioxide levels will have on global temperatures has been raised by the realisation that some positive feedback loops may increase the rate at which temperature increases. However, there are also several negative feedback loops that may slow down the rate of temperature increase.

(a) Explain how the melting of ice from Greenland or the Himalayas is an example of a positive feedback loop.

(b) Explain how the effect on plants of increasing temperature and carbon dioxide levels may be an example of a negative feedback loop.

6 The growing of biofuels has been suggested as one way in which global warming might be slowed down.

(a) Explain how the use of biofuels can reduce overall carbon dioxide emissions.

(b) Evaluate whether this strategy could be used on a large scale to slow global warming.

UNIT 2

Module 2 Chemical processes in the atmosphere

Introduction

In this module you will look at three examples of how the action of human beings has affected processes in the atmosphere. In each case there are considerable environmental implications – in two of these cases, however, the combination of new technology and international political cooperation has provided at least a partial solution. Along the way you will learn a lot more about some important types of chemical processes.

First you will study the issue of acid deposition, which is one of the problems caused by the burning of fossil fuels. In the process, you will learn more about the nature of acids and the process of oxidation, which is involved in their formation. Chemical equations are used to help to describe these processes. The practical technique of titration – used to measure the amount of acid present in a solution – is introduced. Finally you will look at what has been done over the past twenty years to reduce the acid deposition problem.

You will see the importance of ozone, a stratospheric gas, to life at the surface of the Earth. The well-known story of how the man-made CFC gases have depleted the ozone 'layer' is told again – in the process you will find out more about the chemistry of catalysts and radicals.

Finally we continue our exploration of the processes that contribute to global warming. We look at how greenhouse gases such as carbon dioxide and methane can prevent the escape of energy to space, and see how chemists make use of the same effect to identify the chemical bonds in molecules.

Module contents

1. Acids
2. Chemical equations and the formation of acid rain
3. Clearly a problem
4. Oxidation and neutralisation
5. Solutions to acid deposition
6. Ozone and catalysts
7. The effect of CFCs on the ozone layer
8. The greenhouse effect
9. Greenhouse gases

How science works

During this module you will be covering some of the aspects of How Science Works. In particular you will be studying material which may be assessed for:

- HSW 5a: Carry out experimental and investigative activities, including appropriate risk management, in a range of contexts.
- HSW 5b: Analyse and interpret data to provide evidence.
- HSW 6a: Consider applications and implications of science and appreciate their associated benefits and risks.
- HSW 7c: Appreciate the ways in which society uses science to inform decision-making.

Examples of this material include:

- Discuss the causes and effects of acid deposition (spreads 2.2.3 and 2.2.5).
- Discuss strategies used to minimise damage from acid deposition (spread 2.2.5).
- Use titrations to compare the relative amounts of acid in a solution (spread 2.2.4).
- Explain the role of radicals (such as Cl atoms) in removing ozone (spread 2.2.7).
- Interpret data to explain how CFCs have been replaced by other less damaging molecules.

Test yourself

1. Give some names of acids. How would you test a solution to see if it is acidic?
2. 'The chemical formula of a sulfur dioxide molecule is SO_2'. Explain what you understand by this statement.
3. What do you understand by the term 'oxidation'? Give an example of a simple process which could be described as oxidation.
4. Why do chemical equations have to be 'balanced'? Is this equation balanced: $NO + O_2 \rightarrow NO_2$? If not, how would you balance it?
5. Where is the stratosphere found? What gases make up the stratosphere?
6. What do catalysts do to chemical reactions?
7. Name some greenhouse gases. Why does the presence of greenhouse gases cause global warming?

2.2 (1) Acids

The lakes of Norway and Sweden are set in one of the wildest and most remote ecosystems in Europe. Yet in the 1980s it became clear that the environmental impact of human activities was creating a devastating effect on these pristine wildernesses. There were dramatic changes in the type of algae present in the lakes, and the fish and other animal life in the lakes were dying.

The problem was a rapid increase in acidic substances in the lakes. What were these substances and how had they reached such isolated places? The case study activity in spread 2.2.3 *Clearly a problem* will help you to find out more about the background to this story.

Figure 1 Map of northwestern Europe showing the location of acidified lakes in Norway and Sweden

Acid

A substance which produces H^+ ions in solution.

Ion

An atom or group of atoms with a charge (+ or –) e.g. H^+ and NO_3^-.

Molecule

A particle made up of several atoms bonded together e.g. HCl and SO_2.

Strong acid

One that breaks up into ions completely – it is said to be fully ionised.

Weak acid

One that breaks up into ions partially (only a small fraction of the molecules break up) – it is said to be partially ionised.

What is an acid?

Acids are familiar substances in the chemical laboratory – and it was some of these familiar substances that were found to be responsible for the environmental damage in the Scandinavian lakes. Some examples of acids are shown in Table 1.

Name of acid	Formula of acid	Equation for acid breaking apart into ions
Hydrochloric acid	HCl	$HCl \rightarrow H^+ + Cl^-$
Nitric acid	HNO_3	$HNO_3 \rightarrow H^+ + NO_3^-$
Sulfuric acid	H_2SO_4	$H_2SO_4 \rightarrow 2H^+ + SO_4^{2-}$
Carbonic acid	H_2CO_3	$H_2CO_3 \rightarrow 2H^+ + CO_3^{2-}$

Table 1

- In their pure state, these substances are made up of **molecules** but when dissolved in water they break apart to produce **ions**.
- Hydrochloric acid, nitric acid and sulfuric acid are **strong acids** – carbonic acid is a **weak acid**.

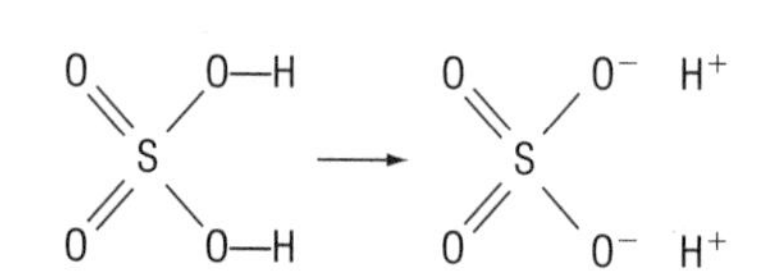

Figure 3 Diagram of the molecular structure of sulfuric acid showing how it breaks apart into ions

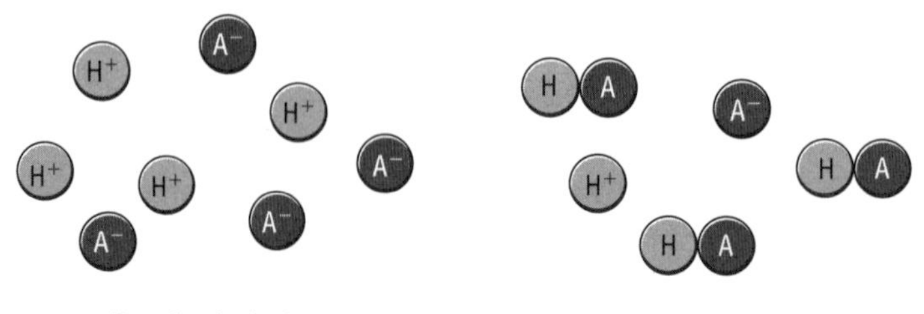

Figure 2 Diagram showing complete and partial ionisation of an acid molecule HA

Properties of acids

You may remember some of these from GCSE.

- Acids have a pH of less than 7.
- Acids are **neutralised** by a range of substances – for example bases, alkalis (bases in solution) and carbonates. The pH of an acidic solution increases when it is neutralised because H^+ ions are removed (they combine with ions from the alkali or base). You will find out more about this in spread 2.2.4 *Oxidation and neutralisation*.

pH

The pH of a solution is a measure of the concentration of H^+ ions – the lower the pH the higher the concentration of H^+ ions.

Mathematically, $pH = -\log_{10}(H^+ \text{ concentration})$ but you will not need to use this relationship in this course.

Concentration

A measure of the ratio of dissolved substance to water in a solution. Often measured in $g\,dm^{-3}$ or $mol\,dm^{-3}$.

Applications: measuring pH

You will have come across the use of universal indicator (in paper or solution) as a way of measuring pH. It is much more accurate to use a pH meter (sometime called a pH probe) to obtain a value for the pH of a solution. It relies on the fact that small voltages can be generated from solutions containing H^+ ions – the size of the voltage depends on the concentration of H^+ ions.

Activity

Use pH meters or indicator paper to measure the pH of solutions of acids. You should try to include sulfuric acid, nitric acid and carbonic acid (a solution of carbon dioxide).

Acid pollutants

Scientists now know that two main types of pollutants are responsible for the formation of acids in the lakes. These are oxides of sulfur (SO_x) and oxides of nitrogen (NO_x) as shown in Table 2.

Type of pollutant gas	Examples	How formed
SO_x	Sulfur dioxide SO_2 Sulfur trioxide SO_3	Burning of fossil fuels which contain sulfur impurities Extraction of metals from sulfide ores
NO_x	Nitrogen monoxide NO Nitrogen dioxide NO_2	From nitrogen and oxygen in the air reacting together in the high temperature of vehicle engines From the burning of fossil fuels containing nitrogen impurities

Table 2

Questions

1 Phosphoric acid (H_3PO_4) is a strong acid, whereas nitrous acid (HNO_2) is a weak acid.
  (a) Give another example of a strong acid and a weak acid.
  (b) Acids ionise (break apart into ions) when they dissolve in water. Give the formula of an ion which is formed when all acids ionise.
  (c) Complete these equations to show what happens when nitrous acid and phosphoric acid ionise:
    (i) $HNO_2 \rightarrow$
    (ii) $H_3PO_4 \rightarrow$

2 **(a)** Name a type of substance which can be used to neutralise an acid.
  (b) Explain why the pH changes when acids are neutralised.

3 Sulfur dioxide (SO_2) and carbon dioxide (CO_2) are both formed when fossil fuels, such as coal, burn in air. Both of these substances can eventually form acids in solution.
  (a) Explain why burning fossil fuels forms sulfur dioxide.
  (b) (i) Suggest the names of acids which might be formed when these gases dissolve in water.
    (ii) Use your answers to part **(b)(i)** to suggest why the release of sulfur dioxide could have a much greater effect on the pH of lakes than the release of carbon dioxide.

2.2  Chemical equations and the formation of acid rain

Chemical equations

Chemical equations are a shorthand way of showing the numbers of particles reacting in a chemical process. They are a helpful way of visualising what happens in a process, as well as an essential way of calculating how much of each substance will be required for a reaction to occur.

You will need to be able to describe the story told by a balanced equation, as well as add numbers to an unbalanced equation.

The key point about a balanced chemical equation is that the number of atoms of each type must be the same on both sides of the equation.

Examiner tip

If you are trying to balance an equation, never ever change the formulas you are given!

Many reactions involving oxygen molecules, O_2, may be easier to balance using a '½' for the number of oxygen molecules. Some people feel uncomfortable about this but it is perfectly acceptable because the equation is really just giving information about the *ratio* of the numbers of particles reacting.

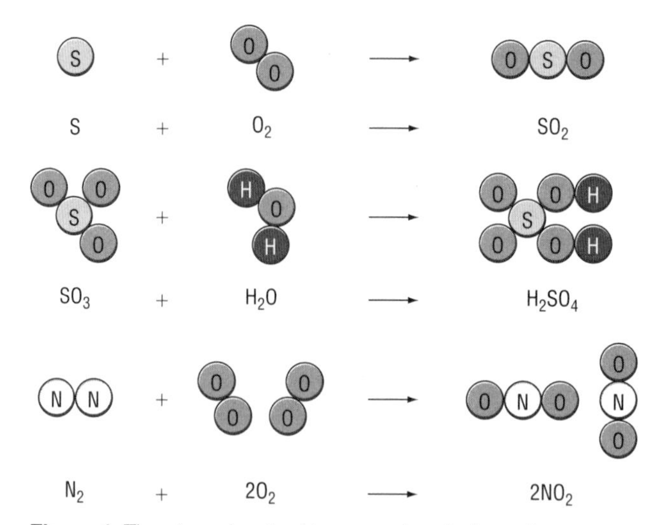

Figure 1 The atoms involved in some chemical reactions

Worked example: balancing equations

$N_2 \quad + \quad O_2 \quad \rightarrow \quad NO_2$

In the unbalanced equation:

2N 2O 1N 2O

So, the N atoms are not balanced – there must be $2NO_2$ molecules to ensure that these atoms are balanced.

Check the balancing again:

$N_2 \quad + \quad O_2 \quad \rightarrow \quad 2NO_2$

2N 2O 2N 4O

Now the O atoms are not balanced – there must be $2O_2$ molecules to ensure that these atoms are balanced.

Check the balancing again:

$N_2 \quad + \quad 2O_2 \quad \rightarrow \quad 2NO_2$

2N 4O 2N 4O

The formation of acid rain

The chemical processes in Figure 2 are all involved in the formation of acid rain. Some of them show how pollutant gases are formed in factories and vehicles, others show how these pollutant gases are converted into acidic solutions, which then fall as acid rain, or as snow or fog – which explains why acid rain is more correctly termed 'acid deposition'.

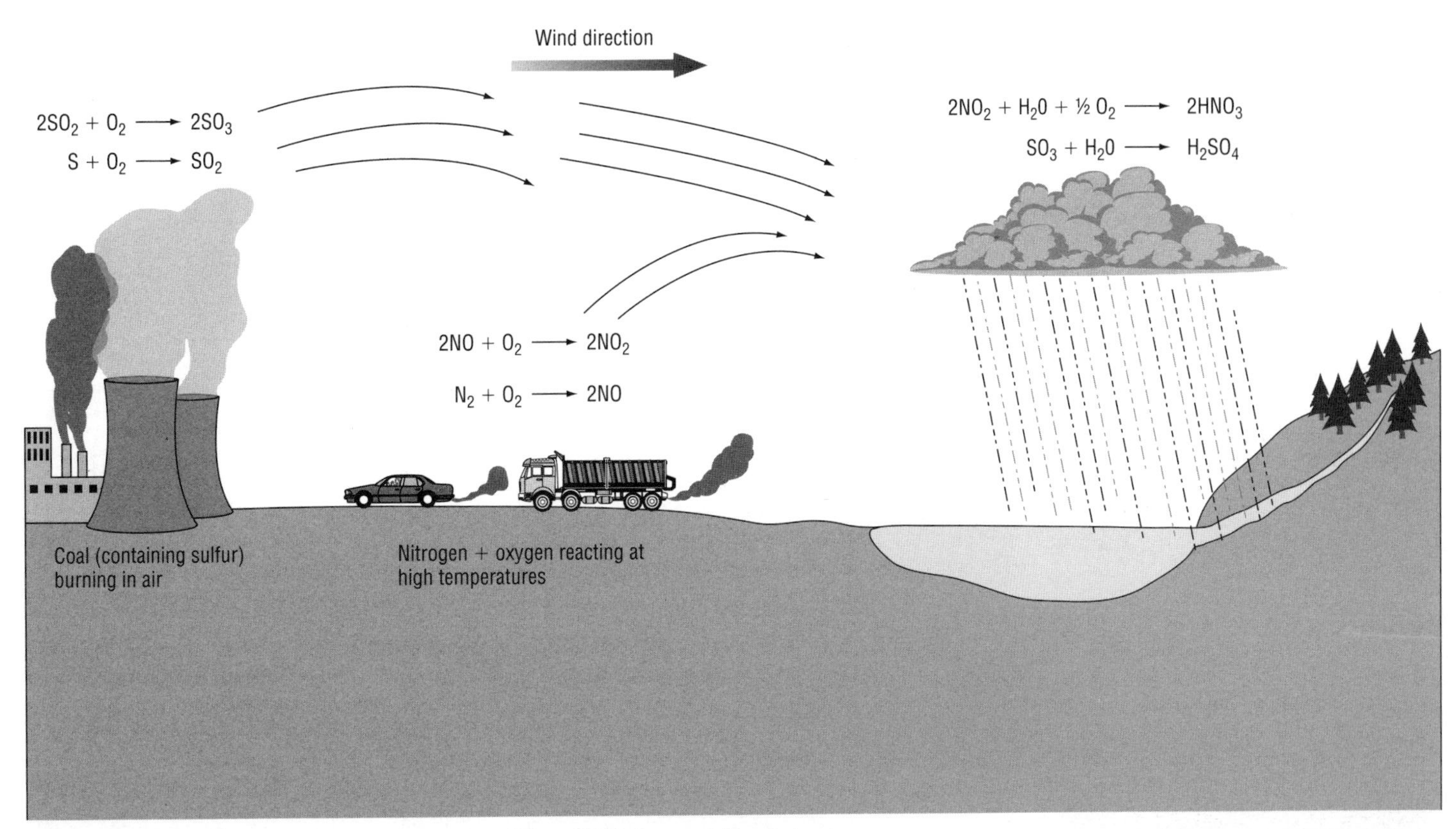

Figure 2 The chemical processes occurring when SO_x and NO_x form acidic substances

Questions

1 Name three different types of acid deposition.

2 Look at the following equations and decide if they are balanced:
  (a) $Mg + H_2O \rightarrow MgO + H_2$
  (b) $C + Cl_2 \rightarrow CCl_4$
  (c) $Na + O_2 \rightarrow Na_2O$
  (d) $CaCO_3 \rightarrow CaO + CO_2$

3 The following reactions are involved in the formation of pollutant gases. Write appropriate numbers in front of the formulas to balance the equations:
  (a) $NO + O_2 \rightarrow NO_2$
  (b) $CO_2 + C \rightarrow CO$
  (c) $SO_2 + O_2 \rightarrow SO_3$
  (d) $N_2 + O_2 \rightarrow N_2O$

4 The following equations are correctly balanced, but one formula is missing in each case. Deduce the formula of the missing substance:
  (a) $NaOH + HCl \rightarrow$ ____ $+ H_2O$
  (b) ____ $+ H_2O \rightarrow H_2SO_4$
  (c) ____ $+ O_2 \rightarrow 2NO$
  (d) $2NO_2 + ½O_2 +$ ____ $\rightarrow 2HNO_3$

5 The effects of acid pollutants are often observed many hundreds of miles away from the source of the acid gases. Explain why.

Examiner tip

You may be asked to balance reactions which relate to the formation of acid pollutants and acid rain. However, you may need to be able to understand a wider range of chemical equations, so some other types of reactions are given in the questions in this spread.

Clearly a problem

How science works

In this spread you will examine and interpret data which provide evidence for the causes of acid deposition (HSW 5b).

Store Hovvatn means 'the lake where the Vikings worshipped'. Like thousands of other lakes in Scandinavia, it looks very beautiful – the water is crystal clear. But the beauty is deceptive – dive in and you find yourself in a strange and eerie world. There are no fish and the water is clear because the free-floating algae – the minute plants which normally float in the water – have disappeared. They have been replaced by a filamentous green alga which has spread a slimy coat over the lake bed and the remains of other plants. In Store Hovvatn, as in many other Scandinavian lakes, the ecological balance has changed during this century. The most striking sign came in the 1960s when the fish populations declined – fish failed to reproduce and mature fish died.

Figure 1 Scandinavian lake

By 1980, Scandinavian scientists had found evidence to suggest that the fish had died because the lakes had become acidic. The most likely cause of this was thought to be acidic pollutant gases dissolved in the rain which fell on southern Scandinavia. As it found its way into the lakes, the acidic rain was thought to damage the fish and kill off the organisms on which they fed.

The commonest acidic pollutant gases in the atmosphere are sulfur dioxide and nitrogen dioxide. Both are produced from the combustion of fossil fuels – both coal and oil contain sulfur compounds as impurities and these produce sulfur dioxide when they burn. Nitrogen dioxide is formed in the heat of the combustion process, particularly in the engines of cars and lorries, when nitrogen and oxygen from the air react. Both sulfur dioxide and nitrogen dioxide are soluble in water – they dissolve in clouds and groundwater to create acidic solutions.

Norway itself produces only small quantities of sulfur dioxide and nitrogen dioxide and the extensive damage was blamed on 'imported' pollution, particularly that blown across from power stations in the UK.

Other possible causes were suggested – for example acidification could be a naturally occurring process; the lake was created at the end of the Ice Age about 10 000 years ago and the conditions changed over time as the lake became more 'mature'. A second suggestion was that a change in land use could be responsible for the increased acidity; some lakes in Scandinavia have become surrounded with conifer forest, in other cases grazing land had gradually become converted to unused moorland.

To investigate the causes of lake acidification, scientists needed reliable data stretching back over hundreds or thousands of years. Direct observation of the land around the lakes, from written records, is only available for the recent past and is very patchy – instead proxy evidence was used from biological indicators in the lake sediments. These indicators – pollen grains and diatoms – allow scientists to deduce information about vegetation and pH.

Long, continuous cylindrical samples of lake sediments are obtained by boring into the floor of the lake. They are then analysed at different depths for key types of pollen and diatoms.

Pollen grains from the plants which grow around a lake are carried into the water, where they sink to become part of the sediment. Pollen grains are resistant to attack by both water and acid, so the history of the vegetation in a lake's catchment area can be deduced from the proportions of pollen grains from certain key species of plant. Diatoms, which are microscopic plants, live in lake water. There are many different types of diatoms – some thrive in water which is close to neutral, some in acidic conditions and others in alkaline conditions. Again, because diatoms are resistant to attack by water and acid, the proportions of the different classes of diatom at different depths in the sediment provide measures of the pH of the lake water in the past.

To trace the history of a lake's pH and surrounding vegetation it is necessary to establish when each layer was laid down. The relative ages of the sediment layers is easy to deduce – the deeper the sediment the greater its age. To get an accurate figure for the absolute age of the sediment, a method that makes use of the presence of naturally occurring radioactive isotopes contained in minerals and organic matter in the lake sediments is used.

The scientists who analysed the lake sediments were able to deduce that acidification of the lakes began about 150 years ago and has taken place ever since. This has occurred in areas where there has been no afforestation or change in land use. It now seems certain that the burning of fossil fuels since the Industrial Revolution is the cause of damage in lakes like Store Hovvatn.

Questions

1. The lake described in this spread has become polluted. Describe the two main environmental effects of this pollution.
2. Scientists have suggested three possible hypotheses for the acidification of the lake. State these three hypotheses.
3. Give the names of the two commonest acidic pollutant gases.
4. Each of these gases is produced from a combustion process. Combustion is the reaction which occurs when a fuel reacts with oxygen from the air. State, as precisely as possible, the source of the sulfur and nitrogen which combine with oxygen in these combustion reactions.
5. In order to study lake acidification, scientists need to obtain proxy evidence about the levels of pH in the lake. Suggest the meaning of 'proxy evidence'.
6. Which type of organisms give proxy evidence about pH levels in the lake?
7. What else has been found in the sediments which give scientists valuable information about the history of the ecosystem of which the lake is a part?

Figure 2 Electron micrograph of a pollen grain

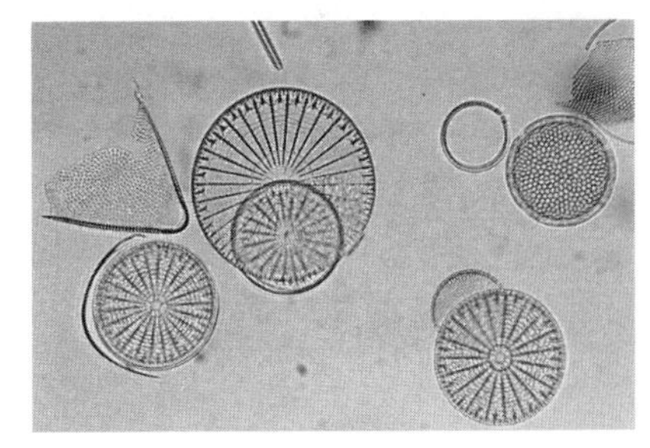

Figure 3 Electron micrograph of a diatom

Other activities

You may be given the opportunity to research some other aspects of acid deposition. These could include:

- a spreadsheet-based activity which will help you to identify the most likely source of the acid pollutants which affected the Norwegian lake
- a practical simulation which helps you to visualise the chemical processes leading to acid deposition
- researching contemporary newspaper and magazine reports to become more aware of the damage that was caused.

Oxidation and neutralisation

Oxidation

You have seen in previous spreads that many of the reactions which cause acidic gases to be released into the atmosphere involve reactions with oxygen. Such reactions are described as **oxidation.** Scientists use the idea of *oxidation state* to identify oxidation processes. These can also be used to identify **reduction** processes.

Calculating oxidation state

In simple reactions such as the formation of acid rain, you will only need to use a few simple rules to calculate a number which tells you the oxidation state of each of the atoms in chemical substances:

- The oxidation state of any uncombined element is 0.
- The total oxidation number of all the atoms in a compound adds up to 0.
- Oxygen atoms when combined in a compound have an oxidation state of –2.
- Hydrogen atoms when combined in a compound have an oxidation state of +1.
- If more than one atom of each type is present, work out the oxidation state of just one atom.

How science works

In this spread you will be introduced to the use of titrations to find the amount of acid in a solution (HSW 5a).

Oxidation

A reaction in which the oxidation state of an atom increases during a chemical process. This occurs when, for example, oxygen is gained by the atom.

Reduction

A reaction in which the oxidation state of an atom decreases during a chemical process. This occurs when, for example, oxygen is lost by the atom or when hydrogen is gained.

Worked examples

In the reaction $S + O_2 \rightarrow S\ O_2$

$$\begin{matrix} S & O_2 & S & O_2 \\ 0 & 0 & +4 & -2 \end{matrix}$$

- On the left-hand side of the equation, both sulfur and oxygen are uncombined elements. So they both have oxidation states of zero.
- On the right-hand side of the equation, there are two oxygen atoms. Each has an oxidation state of –2, making –4 in all. So the sulfur atom must have an oxidation state of +4 (+4 and –4 cancel out to make 0).
- The oxidation state of the sulfur atom has increased so it has been oxidised.

In the reaction $SO_2 + H_2O + ½O_2 \rightarrow H_2SO_4$ it is much harder to be certain what has happened to the sulfur.

- The oxidation state of sulfur in SO_2 is +4.
- In H_2SO_4, the 4 oxygen atoms add up to –8 and the 2 hydrogen atoms add up to +2. So the sulfur atom has an oxidation state of +6.
- During the reaction, the oxidation state of sulfur has increased from +4 to +6. So it has been oxidised.

Neutralisation reactions and titrations

Acids react with compounds which contain ions such as hydroxide (OH^-) or carbonate (CO_3^{2-}). Both these ions react with H^+ ions so they neutralise the acid and increase the pH.

$$H^+ + OH^- \rightarrow H_2O$$
$$2H^+ + CO_3^{2-} \rightarrow H_2O + CO_2$$

Examiner tip

You do not need to remember these equations but you may need to be able to explain the significance of them.

The reaction with carbonate ions (CO_3^{2-}) is important because it provides one of the ways of reducing the effects of acid deposition in lakes and rivers. Calcium carbonate (limestone) is added and the carbonate ions react with the H^+ ions as shown above – this is known as *liming* lakes.

It also explains why some regions are more susceptible to acid deposition than others. If the underlying rock beneath soil contains limestone then the carbonate ions will neutralise the H^+ ions, preventing the soil pH from changing.

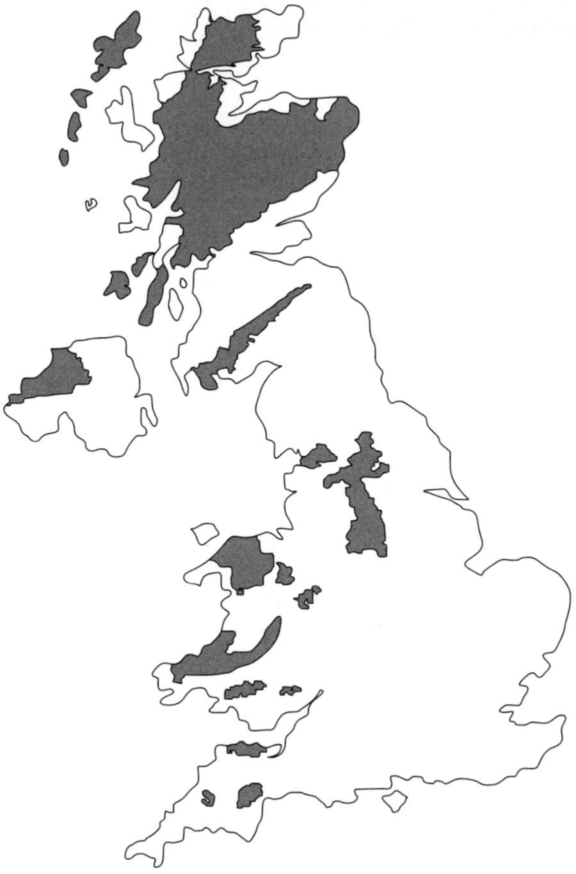

Areas where soil and rocks have little resistance to acidification

Figure 1 Map of the UK showing susceptibility to acid deposition

Titrations

To find the amount of acid present in a solution accurately, scientists use a technique known as an acid–base titration.

Acid solution is measured out into a conical flask accurately using a *volumetric pipette*. A few drops of an indicator are added to the acid in the flask. Then a solution of a base (known as an *alkali*) is added from a *burette*. When the *end point* of the titration is very close, the alkali is added drop-by-drop. At the end point of the titration the indicator suddenly changes colour.

Questions

1. **(a)** If an atom is in the form of an uncombined element, what is its oxidation state?
   (b) What is the oxidation state of **(i)** oxygen atoms in compounds; **(ii)** hydrogen atoms in compounds?
2. In the reaction $N_2 + O_2 \rightarrow 2NO$, the nitrogen and oxygen atoms both change oxidation state.
   (a) What is the oxidation state of **(i)** N in the N_2 molecule; **(ii)** O in the O_2 molecule?
   (b) What is the oxidation state of **(i)** N in NO; **(ii)** O in NO?
   (c) Explain how you can tell by looking at the changes in oxidation states that the nitrogen has been oxidised.
3. Explain why limestone (calcium carbonate, $CaCO_3$) is able to neutralise acids. (Hint: think about what ions are present in calcium carbonate.)
4. 10 cm^3 of a standard solution of nitric acid has a concentration of 0.20 g dm^{-3}. This requires 20 cm^3 of a standard alkali for neutralisation. A 10 cm^3 sample of an unknown acid required 28.7 cm^3 of the standard alkali to neutralise it. Calculate the concentration of the unknown acid.

Examiner tip

You may need to do some calculations from titration data – for example to calculate the concentration of the acid in g dm^{-3}. To do this you will need to compare the titration result with some data from *standard* solutions data (see the worked example below).

Applications: the use of volumetric equipment

Figure 2 Volumetric pipette and burette

These two pieces of equipment are used to enable accurate measurement of volumes. This is because the cross-sectional areas of pipettes and burettes are quite small at the point of measurement – a small change in volume would produce a noticeable change in height.

You may sometimes have to quote the precision of these measuring devices. For example, a burette may have a precision of ± 0.05 cm^3; a 25 cm^3 pipette may have a precision of ± 0.04 cm^3.

Worked example

A standard solution of sulfuric acid has a concentration of 9.80 g dm^{-3}. 25 cm^3 of this solution requires 15.9 cm^3 of a standard alkali for neutralisation.

25 cm^3 of an unknown acid solution requires 9.3 cm^3 of the standard alkali for neutralisation. Calculate the concentration of this unknown acid solution.

Simply use the ratio of the two titration figures – the concentration of the unknown acid is $9.3/15.9 \times 9.80 = 5.73$ g dm^{-3}.

2.2  Solutions to acid deposition

How science works

In this spread you will read about some more of the effects of acid deposition (HSW 6a) and consider strategies for the reduction of acid deposition (HSW 7c).

The problems caused by acid deposition

In the last few spreads you have learnt about some of the problems caused by acid deposition:

- Altering the pH of lakes and rivers, which affects the populations of organisms living in the ecosystem. Some populations die out and others, better adapted, take their place. This has an effect on the whole food web in the lake.
- Corroding stonework, particularly limestone. The limestone (calcium carbonate) reacts with the acid in a neutralisation reaction and the stone dissolves in the acid.

The effect on trees

Trees are also affected badly by acid deposition. The leaves of trees can be damaged by direct contact with acids, particularly in the form of acid fog, which often surrounds the trees in high forests for long periods of time. This causes damage to the surface of the leaves, which may die off as a result.

Most of the damaging effects of acid deposition on trees and other vegetation are due to the effect that the acid has on the soil. Many soils largely consist of small particles of silicate clays. These are negatively charged and have positive ions attached to them. The H^+ ions in the acid, which soaks into the soil, can take the place of these ions. As a result these ions are released into the water contained in the soil. A similar effect occurs in soils with a high humus (organic matter) content – the humus is negatively charged and nutrient ions are normally attached to particles of humus.

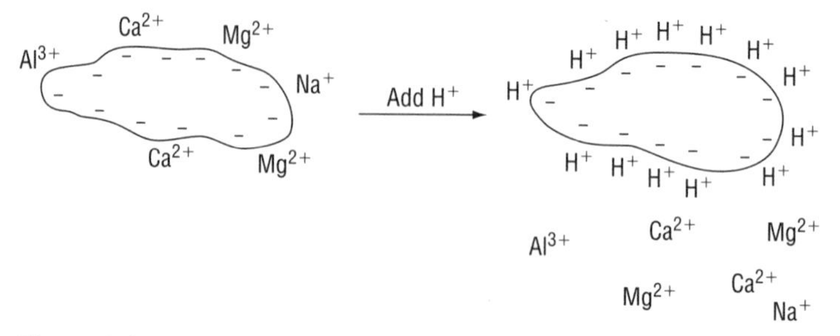

Figure 1 Ion exchange in a clay soil

This process is known as *ion exchange*. It causes two problems for plants growing in the soil:

- Ions such as calcium and magnesium are important nutrients for plant growth. The process of ion exchange means that they are gradually leached out of the soil.
- Aluminium ions are highly toxic to plants. Normally the high charge (3+) on aluminium ions means that they are permanently bonded to the clay particles and cannot cause damage. If they are released by the process of ion exchange then it is possible for plants to absorb them through their roots and they become poisoned.

Minimising the damage from acid deposition

In some ways, the acid deposition problem is an example of how technology has been able to deal with an environmental issue and find effective solutions.

This has involved a great deal of international cooperation. Acid deposition is a cross-boundary issue – the countries which produce most of the acidic gases are not those that suffer most from the effects. This is partly because the effects of acid deposition depend on the soil type (see spread 2.2.4 *Oxidation and neutralisation*) but mostly because acidic gases are carried away from the countries of origin by prevailing winds. In the case of northern Europe this meant that Norway and Sweden received very high levels of acid deposition.

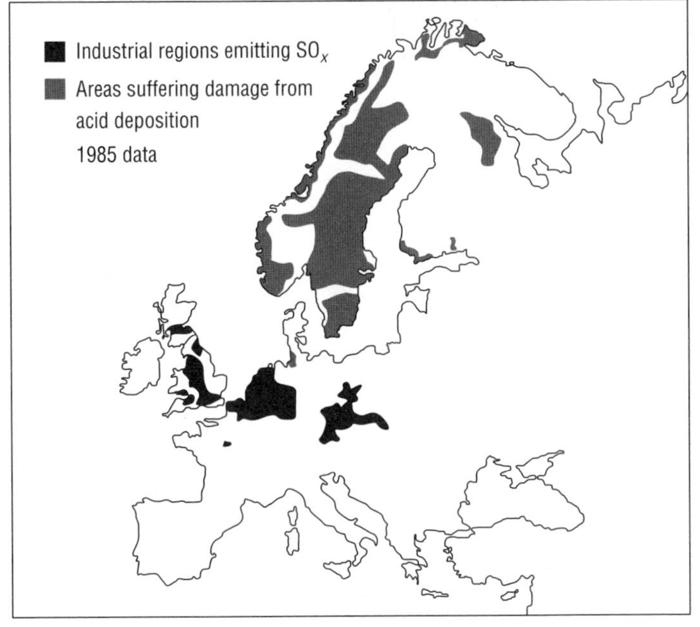

Figure 2 Map of northern Europe showing areas with high SO_2 emissions and areas with high damage due to SO_2 deposition

The Helsinki protocol of 1985 committed European countries to a 30% cut in SO_x emissions. The use of clean technology has allowed many countries to achieve a greater reduction than this, which has alleviated the acid deposition problem in Europe. There are now worries about the increase in emissions from some of the world's developing economies, notably China. To reduce acid emissions even further requires emissions of NO_x to be reduced.

The reduction in SO_x emissions has been achieved by the use of several strategies:

- Desulfurisation of fuels – the sulfur present in fossil fuels can be removed before it is burnt. This has been particularly successful in producing ultralow-sulfur petrol. One method is to react the sulfur atoms with hydrogen: $H_2 + S \rightarrow H_2S$.
- Flue gas desulfurisation – removing SO_x from the mixture of gases which result from the combustion of fossil fuels. This can be done by neutralising the SO_x with bases such as lime (calcium hydroxide) or ammonia. Often the products (such as calcium sulfate) are useful by-products which can be sold.
- Reduction in NO_x formation – this has been tackled by using catalytic converters in car exhaust systems to convert NO_x into harmless nitrogen (N_2).

Applications: UK and Norway

In 1990, Norway produced 66×10^3 tonnes of SO_2 whereas the UK produced 3832×10^3 tonnes. However, total acid deposition in Norway was 288×10^3 tonnes whereas total acid deposition in the UK was 1120×10^3 tonnes.

The implication is that much of the UK's acid emissions are 'exported' to Norway on the prevailing south-westerly winds.

Activity

You can look at data showing how the emissions of acidic gases reaching Norway have changed over the last 20 years – the Norwegian government has a good site providing this data. Search the Internet for 'Norway + acid rain'.

Applications: technological solutions to environmental problems

Clean technology – the use of new technology within a process to minimise emissions which may cause environmental damage.

Clean-up technology – if emissions occur from a process, clean-up technology can be used to minimise the environmental damage they cause. For example, adding lime to lakes to neutralise the acid already present.

Questions

1 Match the effects of acid deposition with the correct explanations for the effects.

Effect	Explanation
Kills trees	Acid reacts with carbonate ions
Kills fish in lakes	Ion exchange releases toxic ions
Corrodes stonework	Change in pH alters food chains

2 Give two reasons why Norway has been particularly badly affected by acid deposition in the past fifty years.

3 What change do you think has occurred in China to make it the biggest emitter of SO_x in the world?

4 Describe one example of clean technology which has been used to reduce **(a)** SO_x emissions; **(b)** NO_x emissions.

2.2  Ozone and catalysts

Ozone

Ozone (O_3) is a gas found mainly in the stratosphere – the layer of the atmosphere which lies above the troposphere. It can be formed from normal atmospheric oxygen in a *photochemical* reaction: $3O_2 \rightarrow 2O_3$.

This reaction is only possible in the stratosphere because it requires short wavelength ultraviolet (UV) radiation. Almost none of this short wavelength UV radiation reaches the troposphere.

Although it makes up only about 1% of the stratosphere, ozone plays a vital role in absorbing slightly longer wavelength UV radiation from the Sun. If this radiation reached the surface of the Earth it would cause tremendous damage to cells, producing mutations and causing cancer.

So the *formation* of ozone, and the ozone molecules themselves, shield the Earth's surface from a range of UV radiation.

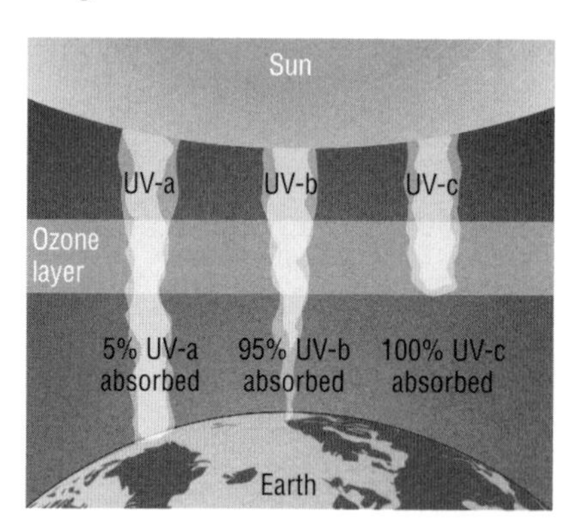

Figure 1 Absorption of UV radiation by ozone

Examiner tip

You won't be expected to remember the whole of this series of reactions. But you should be able to see that these equations balance, and so be able to complete an equation where a 'gap' has been left for a missing substance.

Catalyst

A substance which speeds up the rate of a chemical reaction without being used up – it provides a new pathway with a lower activation energy.

Removal of ozone

The amount of ozone in the atmosphere will normally stay constant because ozone is continually being removed from the atmosphere by natural processes. One of these involves nitrogen monoxide, NO, and oxygen atoms, O. These are both present in small amounts in the stratosphere.

The reaction happens in two steps:

Step 1: $O_3 + NO \rightarrow NO_2 + O_2$

Step 2: $NO_2 + O \rightarrow NO + O_2$

Overall: $O_3 + \cancel{NO} + \cancel{NO_2} + O \rightarrow \cancel{NO_2} + \cancel{NO} + O_2 + O_2$

$$O_3 + O \rightarrow O_2 + O_2$$

You should be able to see how these two equations have been added together and simplified to show how simple the overall equation is. Because the NO is not actually used up in the reaction (it is regenerated in the second step) it is described as a **catalyst.** The two steps shown here are a *catalytic cycle*, which can be repeated over and over again with just small amounts of NO molecules.

How do catalysts work?

The key to this is to look at what happens in reactions such as the breakdown of ozone.

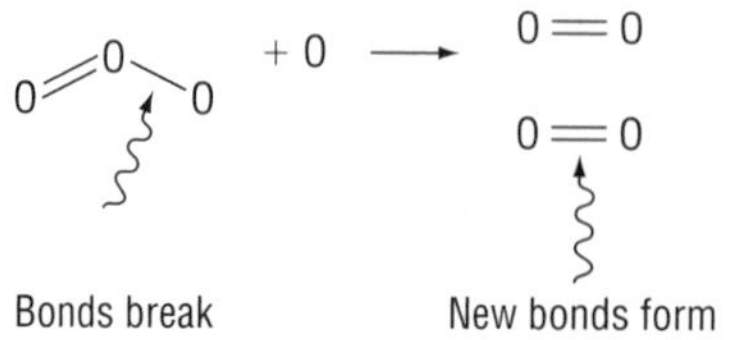

$O{=}O{-}O + O \rightarrow O{=}O$ + $O{=}O$

Bonds break

New bonds form

Figure 2 Bond breaking and forming in $O_3 + O \rightarrow 2O_2$

The reaction involves bond breaking (which requires energy) and bond forming (which releases energy) (see spread 2.4.1 for more details). The problem is that the energy required for an oxygen atom to pull one of the oxygen atoms out of the ozone molecule is very large. In fact, it requires so much energy that it just won't happen under normal conditions – we say that the **activation energy** is too high.

But nitrogen monoxide (NO) is much more reactive than oxygen. It can strip an oxygen atom away from the ozone molecule more easily – so the activation energy is reduced.

Activation energy

The minimum energy required to break the bonds in a molecule in order that a reaction can proceed (and new bonds form).

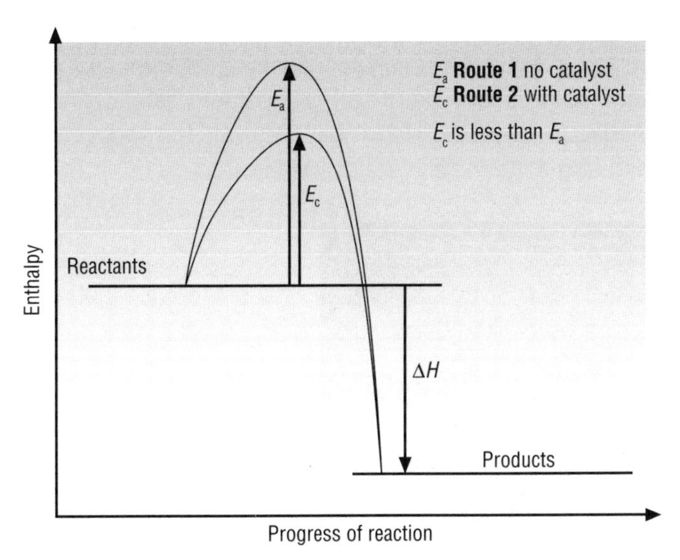

Figure 3 Energy profiles: for a catalysed reaction and for an uncatalysed reaction

Homogeneous and heterogeneous catalysts

The nitrogen monoxide molecule in the ozone removal reaction is in the form of a gas. Ozone is also a gas, and so the nitrogen monoxide is described as a **homogeneous catalyst**.

In spread 2.2.5 you saw that catalytic converters are used to remove NO_x from car exhaust gases. A catalytic converter is a solid structure in which the surface is covered with platinum metal. Gases such as nitrogen monoxide and carbon monoxide bond to the catalyst surface and this makes it easier to break the bonds holding the atoms together, and then to rearrange them into less harmful molecules.

Homogeneous catalyst

A catalyst which is in the same physical state as the other substances in the reaction – for example when all the substances are gases.

Heterogeneous catalyst

A catalyst which is in a different physical state to the other substances in the reaction – for example when the catalyst is solid and the other substances are gases.

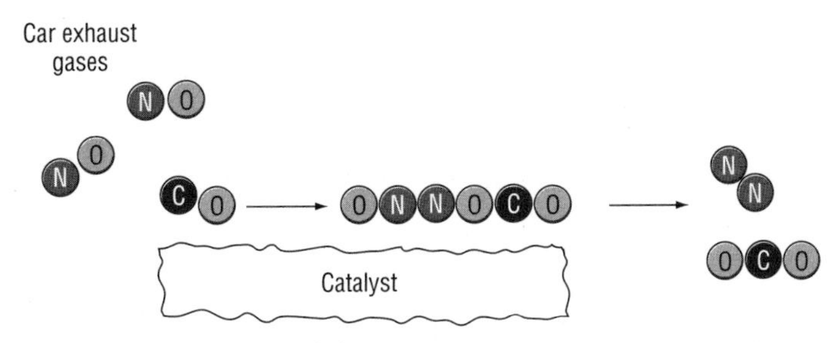

Figure 4 Action of a catalytic converter

Questions

1 Explain why the ozone in the Earth's atmosphere is important for human health.
2 Explain why we can say that nitrogen monoxide, NO, acts as a catalyst for the breakdown of ozone.
3 Briefly describe how catalysts work.

2.2  The effect of CFCs on the ozone layer

How science works

In this spread you will examine how the use of man-made CFC molecules had an unexpected effect on the ozone in the stratosphere (HSW 6a). You will also consider how CFCs can be replaced by other less damaging molecules (HSW 7c).

As you saw in spread 2.2.6, the amount of ozone in the stratosphere is kept constant – in a steady state – because the rate at which it is being formed is equal to the rate at which it is being removed. However, during the 1980s there was sudden concern from the world's atmospheric scientists because in certain parts of the world – particularly over the polar regions at the beginning of spring – the amount of ozone in the stratosphere seemed to be dropping alarmingly. This was the finding that caused the public to start talking about the 'hole in the ozone layer'.

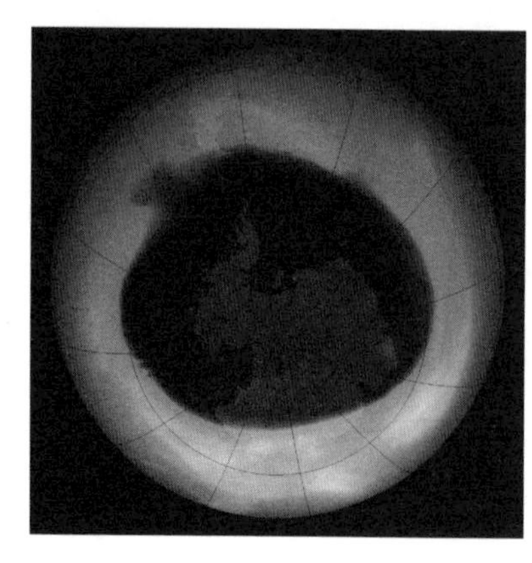

Figure 1 Satellite data showing the extent of the ozone 'hole' in 2005. The dark area shows where ozone concentration is very low

The cause of ozone depletion – CFCs

One possibility was the increased amount of NO in the atmosphere from aircraft emissions, or from the oxidation of dinitrogen oxide resulting from fertiliser use in the soil:

$$2N_2O + O_2 \rightarrow 4NO$$

However, in the 1980s it became clear that the main cause was the release of man-made chemicals called chlorofluorocarbons (CFCs) into the atmosphere.

CFCs include molecules such as CFC-11 (CCl_3F) and CFC-113a (CF_3CCl_3).

CFCs became important in the 1950s because – being unreactive, non-toxic, stable liquids – they could be used as refrigerant fluids in fridges and air-conditioners and as propellants in spray cans.

Chlorofluorocarbons

Small organic molecules which contain chlorine and fluorine atoms bonded to carbons atoms.

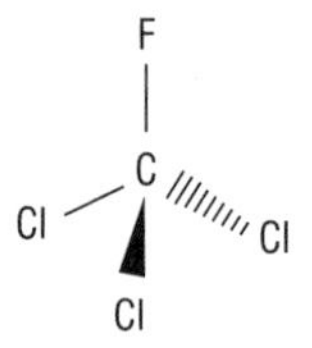

Figure 2 Shape of a CFC-11 molecule

How CFCs deplete ozone

CFCs are very stable molecules and this means that they do not break down in the troposphere. Instead, they diffuse into the stratosphere where they are exposed to ultraviolet radiation. This breaks the bonds between carbon atoms and chlorine atoms in the molecules.

$$CCl_3F \rightarrow CCl_2F + Cl$$

Figure 3 CCl_3F breaking down to produce a Cl atom

The chlorine atom then acts like a NO molecule, and breaks ozone down in a catalytic cycle:

Step 1: $O_3 + Cl \rightarrow ClO + O_2$

Step 2: $ClO + O \rightarrow Cl + O_2$

Overall: $O_3 + O \rightarrow O_2 + O_2$

Radicals

The reason that both NO and Cl act as catalysts, providing a new pathway to break down ozone molecules, is that they are both **radicals**.

Figure 4 Dot-and-cross diagram of NO and Cl radicals, showing the electrons in the outer shell of the atoms

Dealing with the ozone problem

As with the story of the reduction in SO_x emissions in Europe, international cooperation was needed to implement a strategy to limit the effect that man-made pollutants were having on the ozone layer. In 1985 the Montreal protocol was signed which committed the world to phasing out or drastically reducing the manufacture and emission of CFCs and other ozone-depleting substances. More affluent countries even gave economic help to poorer countries to help them to implement the strategy. The policy has been almost completely successful and CFC emissions are now extremely low. However, the CFCs emitted over the past 50 years are still present in the atmosphere and the chlorine atoms formed from them are still causing ozone depletion. Not until the end of the twenty-first century is it likely that ozone depletion due to CFCs will have completely ceased.

Questions

1 Give three possible sources of substances which could cause ozone depletion.

2 State three uses of CFCs which made them important commercial chemicals in the second half of the twentieth century.

3 CFCs are very stable. Which part of the atmosphere are they able to reach as a result of this stability?

4 (a) Complete this equation to show how a CFC could break down to produce a Cl atom:

$CCl_2F_2 \rightarrow$ ____ + Cl

(b) Explain how Cl atoms can cause ozone to break down.

5 H atoms contain 1 electron, N atoms contain 7 electrons and O atoms contain 8 electrons. Which of the following substances could be radicals: NO_2, HO, H_2O, HNO_2, NO?

6 The ozone-depletion potential (ODP) gives an estimate of the possible damage which a substance can do to ozone in the stratosphere. Use the following data to suggest a suitable alternative to the use of CFC-11, which was used as a refrigerant fluid. Explain your reasoning.

Compound	ODP	Boiling temperature/°C	Toxicity	Flammability	Lifetime/years
CFC-11	1.0	24	Low	Low	45
HCFC 141b	1.2	32	Low	High	9.3
HCFC 123	0.02	29	Low	Low	1.3
HCFC 22	0.06	−41	Low	Low	22

[Source: US Environmental Protection Agency]

Examiner tip

You will not need to be able to draw dot-and-cross diagrams yourself, but you should be able to understand what they show and comment on any important features, such as unpaired electrons, lone pairs, bonding pairs etc.

Radical

An atom or molecule with one (or more) unpaired electrons. Most radicals have an odd number of electrons so one of these electrons cannot be part of a pair.

Examiner tip

You may be given data from which you can decide which substances could be used to replace CFCs. In many cases, the key thing will be to find a molecule with a similar boiling temperature but with a lower ODP (ozone-depletion potential). Other factors such as toxicity and flammability may also be important.

2.2 (8) The greenhouse effect

You have seen two examples of environmental issues where international cooperation has begun to tackle the cause of the environmental damage. But in the case of what seems likely to be the biggest of all environmental threats – global warming – such international cooperation seems to be unachievable.

The mechanism of global warming

You saw in spread 2.1.9 that greenhouse gases such as carbon dioxide and methane trap infrared radiation emitted from the Earth. The way in which this happens – the mechanism – is well understood by scientists.

Vibrating bonds

Carbon dioxide has molecules in which a carbon atom is covalently bonded to two oxygen atoms. These carbon–oxygen bonds vibrate, like all covalent bonds.

Figure 1 Structure of carbon dioxide, showing vibrating bonds

All bonds vibrate – the exact frequency at which they vibrate depends on:

- the mass of the atoms in the bond – bonds between heavier atoms vibrate at a lower frequency
- the strength of the covalent bond – strong bonds (such as double bonds) vibrate at a higher frequency.

Some examples of typical frequencies at which various bonds vibrate are shown in Table 1. Scientists often prefer to use wavenumbers to describe the frequency of the vibration.

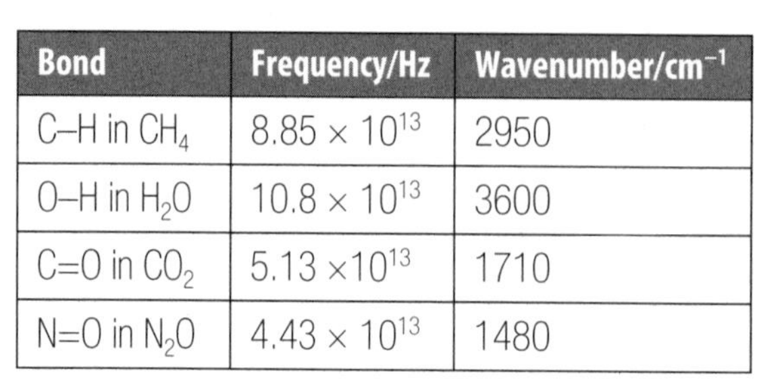

Bond	Frequency/Hz	Wavenumber/cm^{-1}
C–H in CH_4	8.85×10^{13}	2950
O–H in H_2O	10.8×10^{13}	3600
C=O in CO_2	5.13×10^{13}	1710
N=O in N_2O	4.43×10^{13}	1480

Table 1 Frequencies at which some bonds vibrate

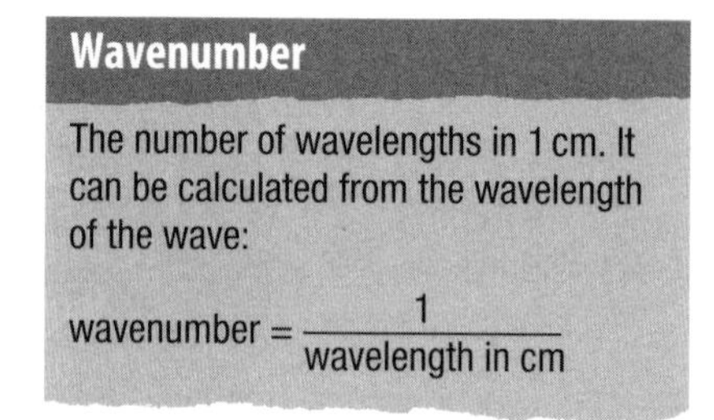

Wavenumber

The number of wavelengths in 1 cm. It can be calculated from the wavelength of the wave:

$$\text{wavenumber} = \frac{1}{\text{wavelength in cm}}$$

Infrared spectroscopy

It is possible to measure how much infrared radiation (IR) is absorbed by substances at different wavenumbers – this is displayed in an IR spectrum.

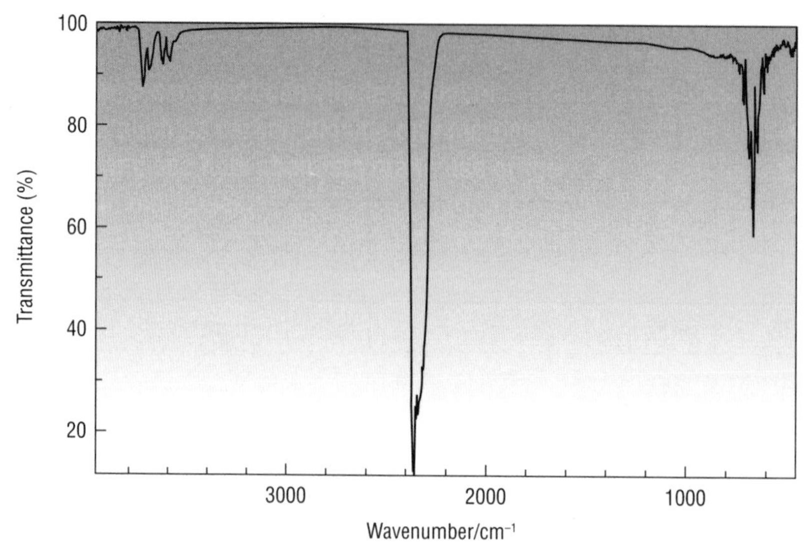

Figure 2 IR spectrum of CO_2

Converting between frequency, wavelength and wavenumber

C–Cl bonds have a wavenumber of 650 cm^{-1}. What frequency does this correspond to? You will need to use the fact that *c*, the velocity of light (and all electromagnetic radiation), is $3.00 \times 10^8 m s^{-1}$.
Wavelength = 1 / 650 = 0.001 54 cm. To convert this to metres you need to divide by 100.
Wavelength = 0.00154 / 100 = 1.54×10^{-5} m.
To calculate the frequency you need to remember that velocity = frequency × wavelength ($c = f\lambda$).
So, frequency = velocity / wavelength = $3.00 \times 10^8 / 1.54 \times 10^{-5} = 1.95 \times 10^{13}$ Hz.

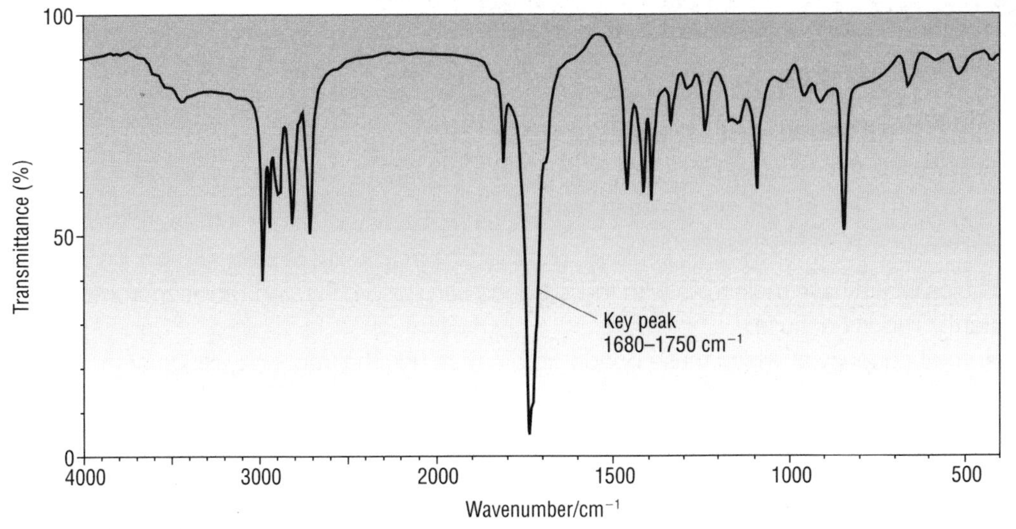

Figure 3 IR spectrum of propanone

Examiner tip

You may be given a table showing data about the wavenumbers at which a variety of bonds absorb infrared radiation. You may then need to identify the bonds present in a molecule from its infrared spectrum. In Figure 3, the absorption 'peaks' at 1700 and 3000 cm^{-1} suggest the presence of C=O and C–H bonds.

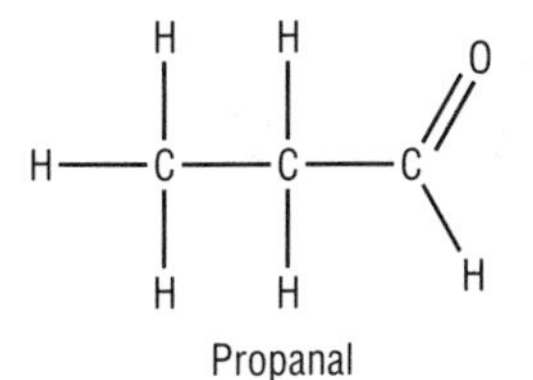

Figure 4 Structural formula of propanal

Questions

1 **(a)** What two factors affect the frequency at which bonds vibrate?

(b) Predict which bond in each of these pairs will vibrate at the highest frequency:

(i) C–H and C–O

(ii) C–O and C=O.

2 A C–O bond vibrates at a wavelength of 9.0×10^{-6} m.

(a) Calculate the frequency of this vibration (use the equation $c = f\lambda$; $c = 3 \times 10^8\,m\,s^{-1}$).

(b) **(i)** What is the wavelength of the vibration in centimetres?

(ii) Calculate the wavenumber of the bond (wavenumber = 1 / wavelength in cm).

3 The infrared spectrum of ethanol, a simple organic compound, is shown in Figure 5.

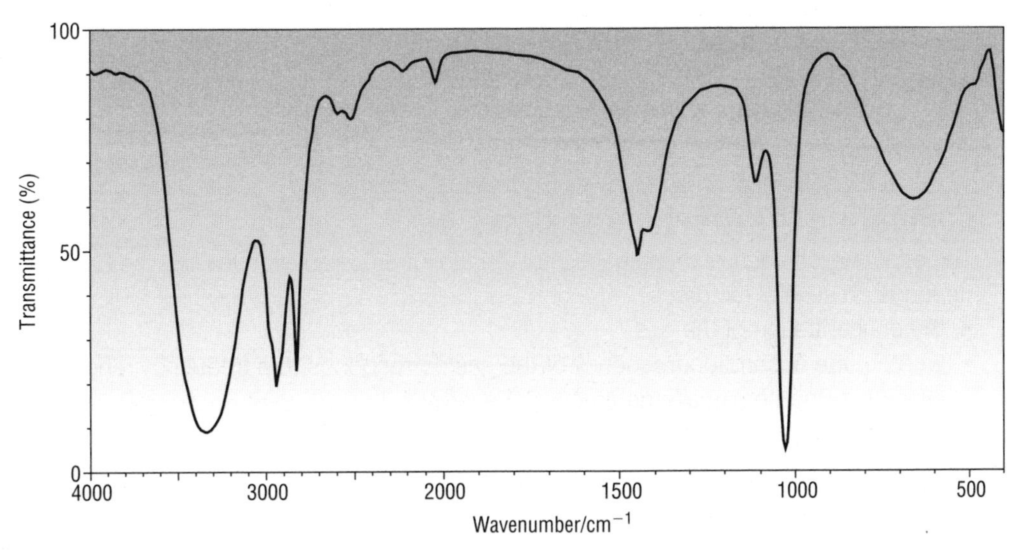

Figure 5 IR spectrum of ethanol

Use this spectrum and Table 1 to identify two bonds present in ethanol. Explain your reasoning.

Greenhouse gases

Which gases are greenhouse gases?

In spread 2.2.8 you saw that C=O, C–H, O–H and N=O all vibrate in the range 4×10^{13} to 11×10^{13} Hz. In fact, bonds often display rather more complex behaviour absorbing in several bands within this region.

Gases which have molecules with these bonds can absorb infrared radiation from the Earth. This is because:

- the frequency at which these bonds vibrate matches the frequency range of the infrared radiation emitted by the Earth
- these bonds are all polar (see spread 2.1.5) – this allows them to interact with the vibrating electric and magnetic field of the infrared radiation.

Greenhouse gases therefore include CO_2, CH_4, H_2O (water vapour) and N_2O.

You can compare the frequency of radiation emitted by the Earth with the frequency at which these gases absorb in Figure 1.

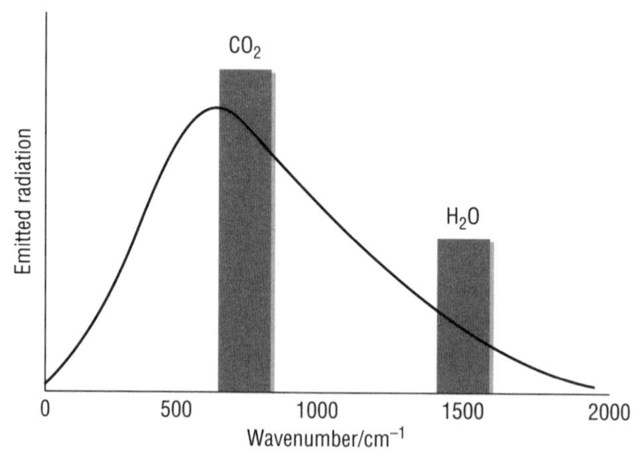

Figure 1 Emission spectrum of the Earth showing the frequency ranges of greenhouse gases

Comparing the greenhouse gases

The effect which each of these gases has on the infrared radiation leaving the Earth depends on several factors:

- the concentration of the gas
- whether the absorption frequency of the gas corresponds to a frequency range in which the Earth emits a large amount of energy
- how efficiently the gas absorbs radiation.

These last two factors can be combined to produce a 'greenhouse factor'. Some data for these greenhouse gases are shown in Table 1.

Gas	Concentration in the troposphere (%)	Greenhouse factor
H_2O	1	0.1
CO_2	0.038	1
CH_4	1.8×10^{-4}	20
N_2O	3.1×10^{-5}	310
CCl_3F (a CFC)	2.6×10^{-8}	3800

Table 1 Concentration and greenhouse factors for the greenhouse gases

Activity

The greenhouse effect has also had an effect on the climate of other planets. Both Mars and Venus have a high percentage of carbon dioxide in their atmospheres – but there are some important differences which mean that the climates have developed in a very different way. Find out about the climate on these planets – try searching the Internet using 'Venus Mars atmosphere'.

Applications: Methane hydrates

Although methane is present in very low concentrations in today's atmosphere, there may have been times when its concentration was much higher and when it may have had a much bigger effect on global climate. One theory is that methane, normally trapped at the bottom of the ocean in structures called methane hydrates (see spread 2.4.7 *Options for future energy generation*), was suddenly released into the atmosphere about 250 million years ago. This could account for the catastrophic change in climate which caused the extinction of almost all marine species at that time.

Water and the greenhouse effect

You will notice that the most abundant greenhouse gas in the atmosphere is water vapour. In fact, we can treat it rather differently – however much water vapour we emit (for example by burning fuels such as natural gas), it will never make a noticeable difference to the amount of water vapour in the atmosphere. The natural processes of evaporation and condensation are what determine the total amount of water vapour in the atmosphere. But these processes could be affected by increasing global temperatures – there will be more evaporation and that would lead to more water vapour in the atmosphere, increasing the greenhouse effect. However, there would probably also be more clouds formed which might reflect more sunlight away from the Earth. So the overall effect of more evaporation is uncertain at the moment – it is another example of the complex feedback loops which climate scientists need to take account of in their models.

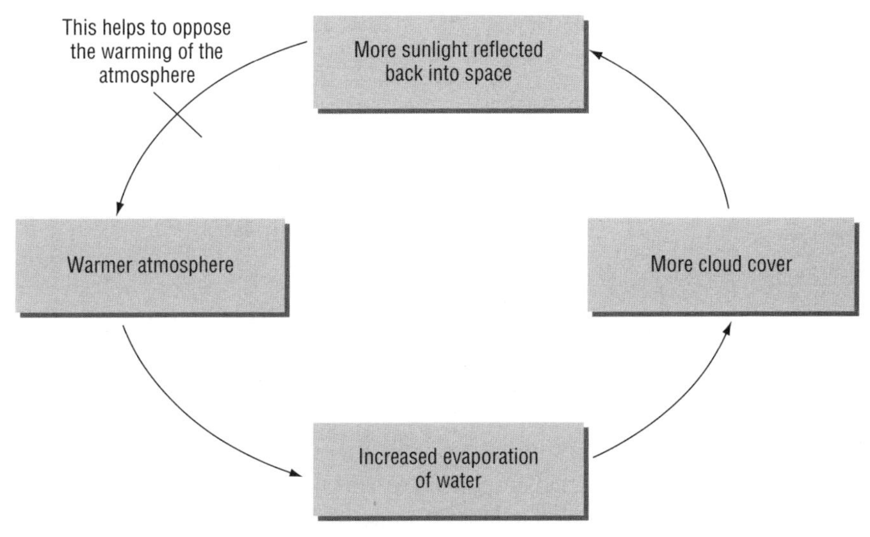

Figure 2 Cloud formation could act as a negative feedback loop in global warming models

Questions

1 Water enters the atmosphere through processes such as evaporation and transpiration from plants. List ways in which the other greenhouse gases enter the atmosphere. How many of these are due to human activity?

2 Use the data in Table 1 to rank the greenhouse gases in order of their overall effect. (Hint: multiply the concentration of the gas by the greenhouse factor.)

3 Draw diagrams to show the structures of the molecules CO_2, H_2O, CH_4 and N_2O. Use the electronegativity values in spread 2.1.5 to identify the polar bonds in these molecules. Use the $\delta+$ and $\delta-$ convention to show the polarity of these bonds.

2.2 Module 2 – Summary

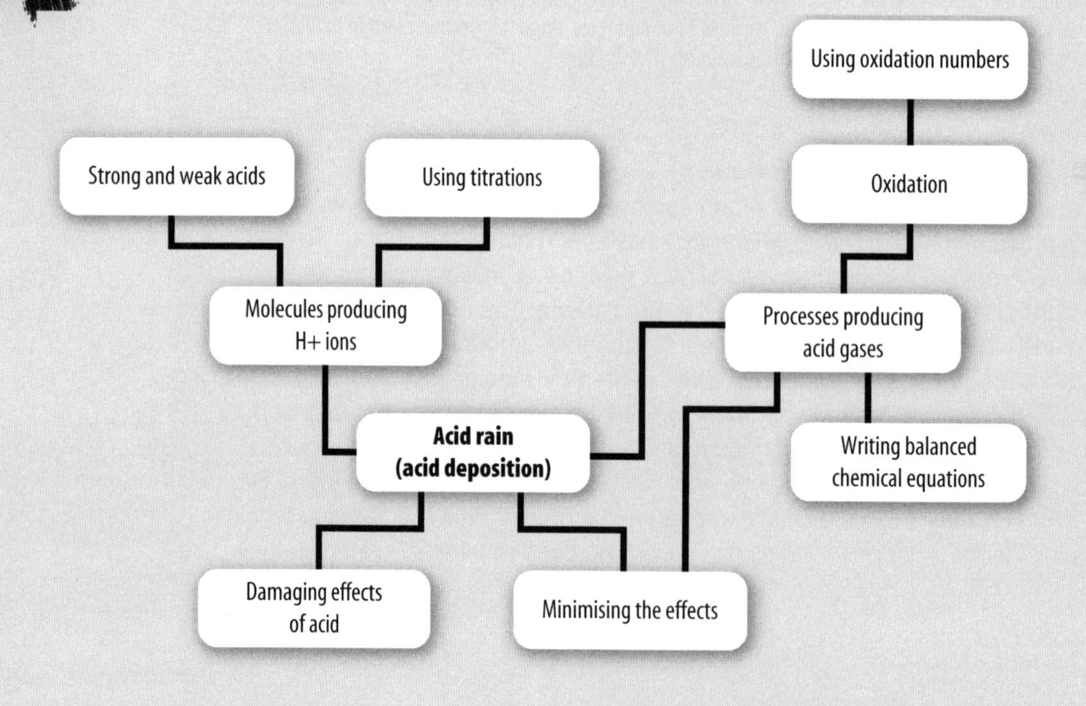

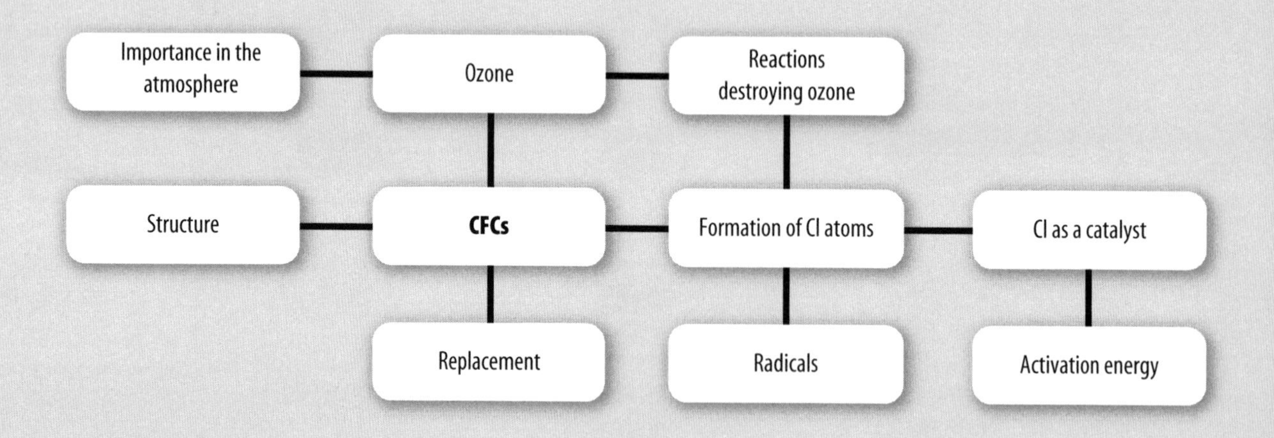

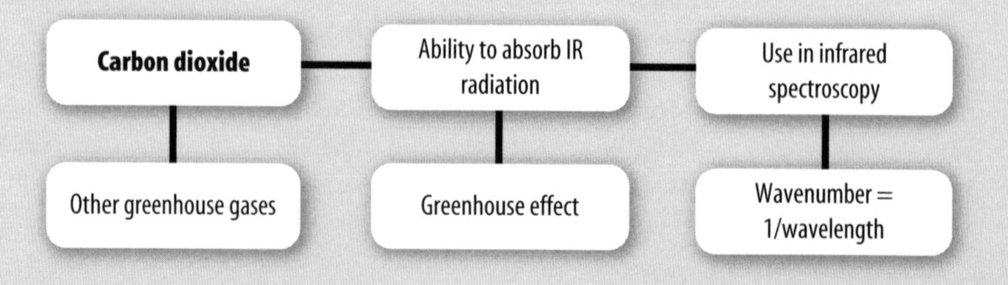

Practice questions

Low demand questions

These are the sort of questions that test your knowledge and understanding at E and E/U level.

1 Sulfur dioxide (SO_2) is one of the pollutant gases which result in the formation of acid rain. One of the ways in which it is formed is from the burning of coal.

(a) (i) Describe how sulfur dioxide is formed when coal burns.

(ii) Suggest a value for the pH of a solution formed when sulfur dioxide dissolves in water.

(b) State two environmental problems which result from acid rain.

2 Ozone (O_3) and carbon dioxide (CO_2) are two gases found in the atmosphere. The table lists some of the properties of these two substances.

	Where found	Type of radiation absorbed	Trend in concentration with time
Ozone		Ultraviolet	Decreasing
Carbon dioxide	Troposphere		

(a) Copy and complete the table.

(b) Give the name of one type of man-made substance responsible for the decreasing concentration of ozone in the atmosphere.

(c) State one major use of the type of substance named in part **(b)**.

Medium demand questions

This is the sort of question that tests your knowledge and understanding at C/D level.

3 Nitrogen dioxide (NO_2) is one of the causes of acid deposition. It is formed in a series of reactions between nitrogen and oxygen. Chemical equations can be written to describe these reactions.

(a) Balance equations 1 and 2, which are unbalanced equations describing the reactions which eventually produce nitrogen dioxide:

Equation 1: $N_2 + O_2 \rightarrow NO$
Equation 2: $NO + O_2 \rightarrow NO_2$

(b) Calculate the oxidation state of N in **(i)** NO; **(ii)** NO_2.

(c) Use your answers to part **(b)** to explain why equation 2 is described as an oxidation process.

(d) Nitrogen oxide (NO) can act as a catalyst for the breakdown of ozone. The two reactions which involve nitrogen oxide are shown in equations 3 and 4:

Equation 3: $O_3 + NO \rightarrow NO_2 + O_2$
Equation 4: $NO_2 + O \rightarrow NO + O_2$

(i) Combine these two equations together to show the overall equation for the breakdown of ozone in the presence of nitrogen oxide.

(ii) Nitrogen oxide is said to be acting as a catalyst. Use equations 3 and 4 to explain how catalysts, such as nitrogen oxide, can speed up reactions.

High demand questions

This is the sort of question that tests your knowledge and understanding at A/B level.

4 CH_4 is described as a greenhouse gas. This means that it can absorb infrared radiation emitted by the Earth. One of the frequencies of infrared radiation absorbed by CH_4 is 8.85×10^{13} Hz.

(a) The frequency (f) and wavelength (λ) of any electromagnetic radiation are related by the equation $c = f\lambda$, where c has the value $3.0 \times 10^8 \, m\, s^{-1}$. Calculate the wavelength (in metres) of the infrared radiation absorbed by methane.

(b) The 'wavenumber' is the number of wavelengths in 1 cm. Show that the wavenumber of the infrared radiation absorbed by methane is approximately 3000 cm^{-1}.

(c) Figure 1 shows a partial infrared spectrum of methane. Copy it and complete the spectrum.

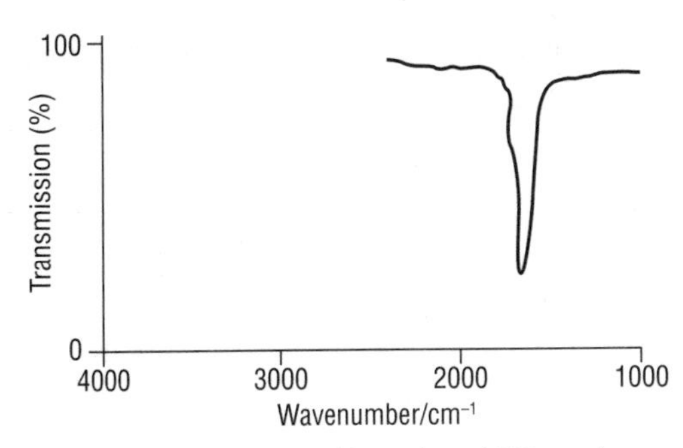

Figure 1 IR spectrum with peak at 1650 cm^{-1}

UNIT 2 Module 3 Proteins and genetic engineering

Introduction

You have already seen, in Unit 1 Module 2, some of the roles which proteins play in cells – for example in the plasma membrane. In this module you take a closer look at these roles – particularly the way in which proteins act as enzymes, biological catalysts. You will see how knowledge of the structure of protein molecules helps to explain how they can take on such a range of complex roles. This knowledge of the bonding and structure also helps to explain why enzymes are so sensitive to the conditions in which they operate.

Then you will look at one of the key breakthroughs in biological science – the understanding of how the structure of the DNA molecules in cells, the genes, can provide a code which eventually determines the structure of all the proteins in the cell, and hence eventually the very characteristics of each living organism.

This knowledge allows scientists to begin to use the techniques of genetic engineering to alter the genes of organisms and to create new organisms with altered and improved characteristics. This has proved enormously controversial, particularly in the way it has been applied to the creation of genetically modified crops. You will look at some applications of this technique and assess the implications of its use.

Module contents

1. **Proteins**
2. **Enzymes**
3. **DNA and genes**
4. **DNA replication and transcription**
5. **The genetic code and protein synthesis**
6. **Genetic engineering**
7. **Concerns about genetically modified crops**

How science works

During this module you will be covering some of the aspects of How Science Works. In particular, you will be studying material which may be assessed for:

- HSW 6a: Consider applications and implications of science and appreciate their associated benefits and risks.
- HSW 6b: Consider ethical issues in the treatment of humans, other organisms and the environment.

Examples of this material include:

- Describe the use of genetic engineering in the development of new crops (spread 2.3.6).
- Discuss the social, ethical and environmental implications of this technique (spread 2.3.7).

Test yourself

1. What are the small molecules from which proteins are built up?
2. Name some of the ways in which proteins are important for the functioning of cells or organisms.
3. What is meant by the term 'catalyst'? How do catalysts work?
4. Where in a cell would you find genes? What do genes do?
5. Why do you think scientists might try to improve a crop using genetic engineering?

2.3 ① Proteins

You have already come across proteins in some of the previous spreads – for example they are present in cell membranes to act as channels which help to control the passage of molecules in and out of cells.

The structure of proteins

Proteins are polymers made up of many smaller units, called amino acids, joined together by peptide links.

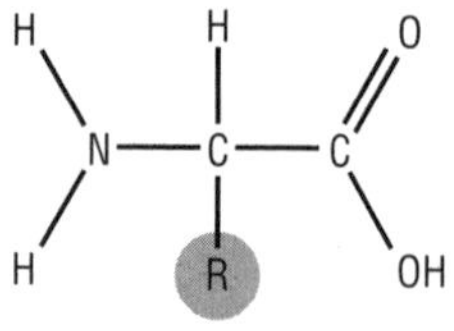

(a) All amino acids have the same basic structure as shown. Only the R group differs between amino acids. Since there are 20 different amino acids, there are 20 different R groups.

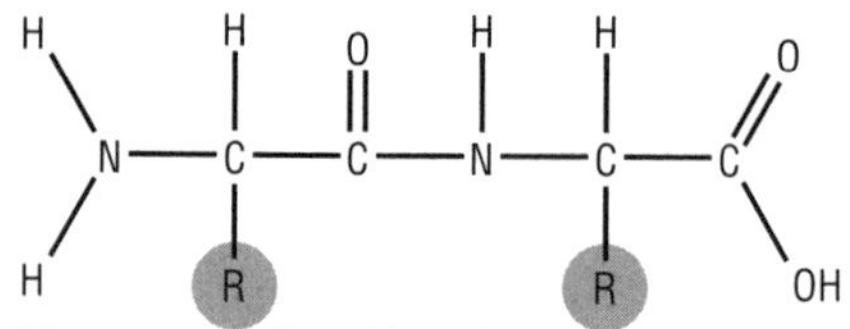

(b) Dipeptide molecule

Figure 1a Amino acid molecule; **b** Two amino acid molecules joined by a peptide link

Examiner tip

You do not need to be able to recall the structure of amino acids, but you should be able to recognise them and features such as the side group and the peptide link between amino acids.

Animals can build up proteins using the amino acids present in food; plants can build up proteins starting from a source of biomass such as glucose, but they also need a source of nitrogen.

There are 20 different amino acids which occur naturally in living organisms. The only difference between the different amino acids is the nature of the side group. These side groups can be important in maintaining the structure of proteins and play an important part in the active site of enzymes.

Levels of structure

Scientists need to study the structure of proteins at many different levels. They may be interested in the way in which the protein is built up from amino acids, or they may be more concerned with the shape of the protein molecule. They talk about three levels of structure:

- primary structure – the sequence of the amino acids in the protein molecule
- secondary structure – the way in which a chain of amino acids is twisted or bent into small structures called an α-helix or a β-sheet
- tertiary structure – the way in which the whole protein is folded into a complex three-dimensional shape.

Figure 2a Primary and **b** secondary structures

Examiner tip

You will not need to be able to draw diagrams such as these, but you will need to be able to recognise features such as secondary structures.

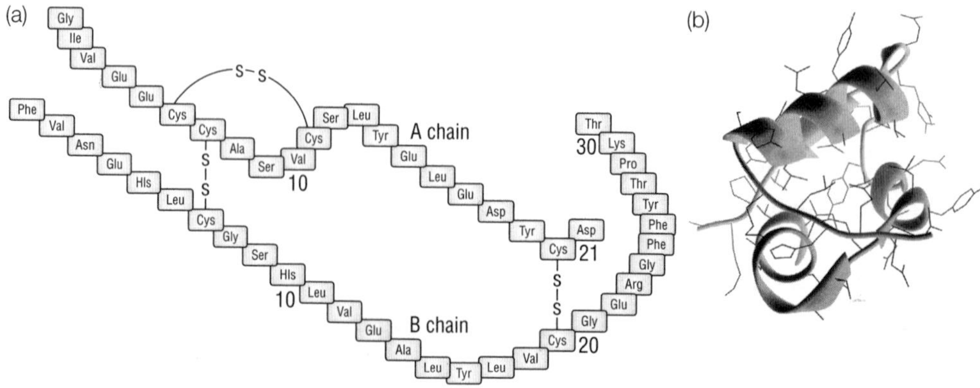

Forces which maintain the structure of proteins

- The primary structure is maintained by the peptide links – covalent bonds which join the amino acid residues together.
- The secondary structure is maintained by hydrogen bonds between, for example, a δ+ H and a δ– O or N atom.
- The tertiary structure is maintained by a variety of forces including hydrogen bonds but also ionic bonds (between charged amino acid groups) and disulfide bridges (between cysteine amino acids which contain S atoms).

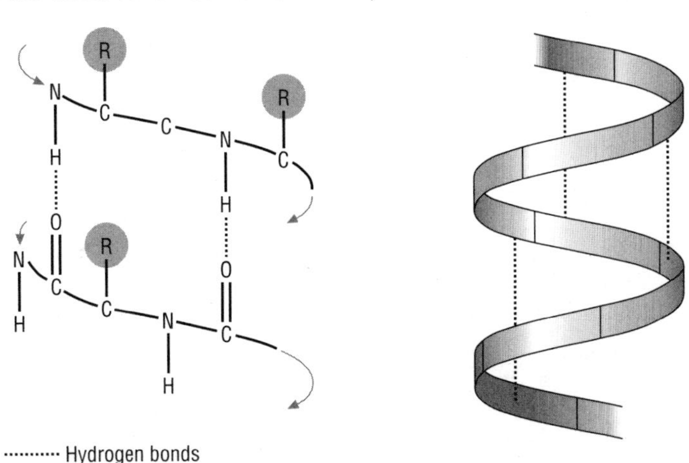

An α-helix has 36 amino acids per 10 turns of the coil. Hydrogen bonds form between one amino acid and the one 'four places' along the chain

Figure 3 Hydrogen bonding between peptide groups in a helix

R-groups sometimes carry a charge, either +ve or −ve. Where oppositely charged amino acids are found close to each other an ionic bond forms

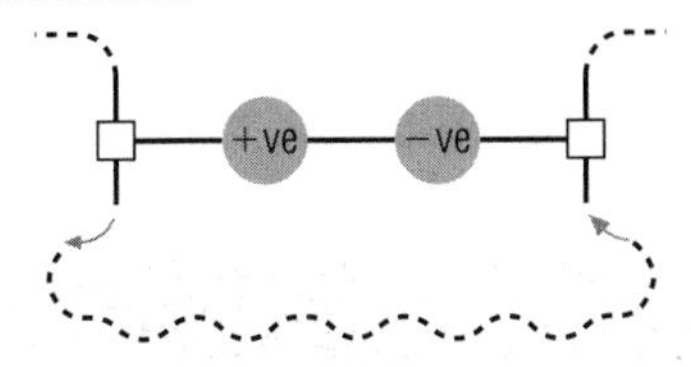

Figure 4 Ionic bonds between ionised amino acid side chains

The amino acid cysteine contains sulfur. Where two cysteines are found close to each other a covalent bond can form

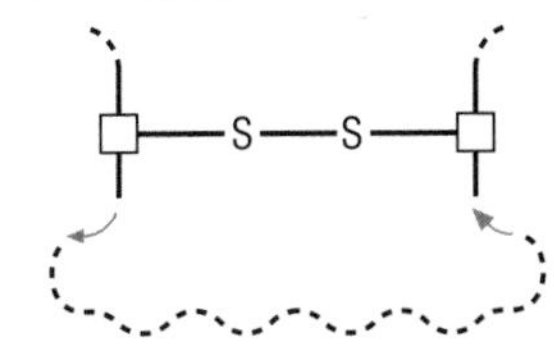

Figure 5 Disulfide bridges

The role of proteins in cells

Because the number of ways in which the 20 different amino acids can be joined together to make protein chains is enormous, there are an almost infinite number of possible protein structures. Different types of structure may have different cellular roles, including:

- channels in membranes – to control the movement of molecules in and out of the cell
- structural molecules – for example making up hair or muscle in animals
- hormones – to regulate the activity of cells
- antibodies – in the immune system
- enzymes – to act as catalysts in biological systems.

Questions

1 Rewrite this table to summarise the meaning of these key words connected with protein structure:

Key word	Meaning
Amino acid	The overall 3D shape of the protein
Peptide link	The twisting of the protein chain into small structural features
Primary structure	The sub-unit from which proteins are built
Secondary structure	The bond which joins amino acids together
Tertiary structure	The sequence of amino acids in a protein chain

2 What type(s) of forces are important in maintaining the following in proteins: **(a)** the primary structure; **(b)** the secondary structure; **(c)** the tertiary structure?

3 Proteins are often described as being necessary for the growth and repair of cells. Give two ways in which proteins are important to *all* cells.

2.3 ② Enzymes

Without enzymes, the complex chemistry necessary for cells to function could not occur. Enzymes enable reactions to happen which would otherwise not be possible, and the highly specific nature of enzymes enables these reactions to be controlled in cells.

Substrate

The molecule on which the enzyme acts – it binds to the active site of the enzyme.

Space-filling diagram

Shows the space taken up by the atoms in a molecule – this shows the tertiary structure of enzymes well.

Ribbon diagram

Shows the protein chain as a ribbon without showing any individual atoms – this shows the secondary structure well.

Enzyme structure

Enzymes have a complex tertiary structure which includes a 'cleft' known as the active site. This is important in explaining the way in which enzymes provide a new pathway for a reaction.

The active site can often be clearly seen in computer-generated images of enzymes.

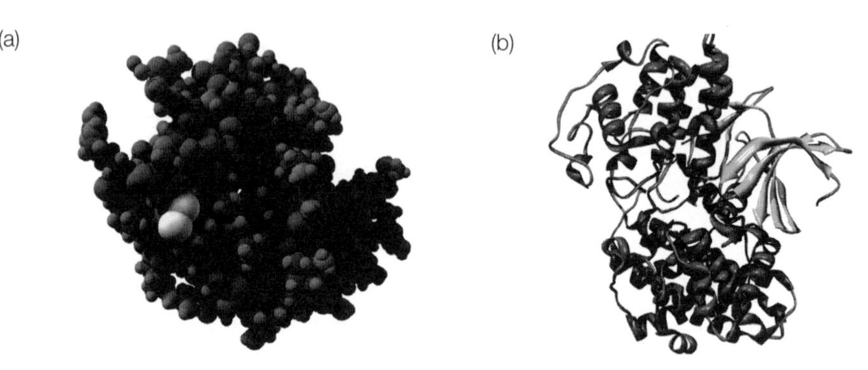

Figure 1 Diagrams of an enzyme: **a** computer-generated space-filling diagram; **b** ribbon diagram

The lock and key mechanism

Enzymes cause substrate molecules to react because they provide a reaction pathway with a lower activation energy. The pathway can be described as a 'lock and key' mechanism:

- The substrate molecule fits into the active site and bonds to it.
- The energy released by the bonds that are made causes the active site to change shape and/or bonds to break in the substrate.
- The substrate reacts to form product molecules.
- The product molecules are released from the active site and diffuse away from it.

This pathway has a lower activation energy than if the substrate molecule was broken down without any enzyme being present.

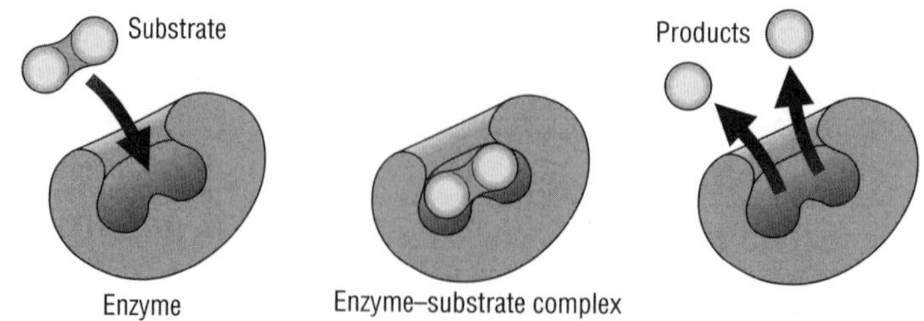

Figure 2 Lock and key mechanism

The substrate bonds to the active site because their structures are complementary. This means that:

- the substrate has exactly the right shape to fit precisely into the active site
- the atoms of the substrate and the active site line up in just the right way to bond together.

These complementary structures explain the **specificity** of the enzyme.

The shape of the active site and the type of atoms at the surface of the active site depend on the side groups of the amino acids present in that part of the protein. A difference of just one amino acid might make the active site have a completely different shape and its action might then be completely different.

Specificity

The word used to describe the fact that only one type of substrate molecule will fit into the active site of an enzyme – so each enzyme will catalyse only one type of reaction.

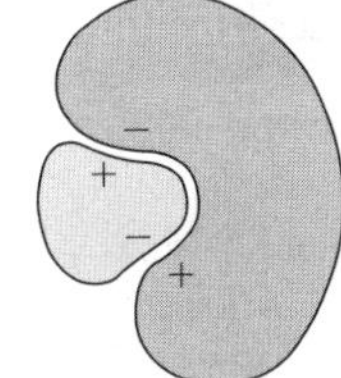

Figure 3 Substrate and active site bonding together

Factors affecting the action of enzymes

The rates of the reactions catalysed by enzymes depend on a number of factors – you may have studied some of these in practical investigations. Many of the patterns can be explained by thinking about how the factors affect the active site:

- Temperature – as with all chemical reactions, temperature affects the rate of an enzyme-catalysed reaction. Increased temperature causes reactions to occur faster, but above a certain temperature (known as the optimum temperature) further heating causes bonds to break and the active site changes shape – the substrate no longer fits and the activity of the enzyme is reduced.
- pH – the concentration of H^+ ions in solutions affects the charges on the side groups in the active site. Each enzyme has an optimum pH at which the charge pattern is just right to maintain the shape of the active site.
- Concentration of substrate and enzyme – increasing the concentration of these substances increases the rate of reaction. However, at very high concentrations of substrate the active site becomes saturated with substrate molecules – all the active sites are full – and so increasing the substrate concentration has no effect on rate.
- The presence of inhibitors – these are molecules which slow down the rate at which enzymes operate. Competitive inhibitors bind to the active site of the enzyme, blocking it and preventing the substrate entering. Non-competitive inhibitors bond elsewhere on the enzyme, but when they do so they cause the active site to change shape.

Applications: the role of enzymes in cells

You have already seen several examples of processes which involve enzymes:

- respiration – enzymes are present in cells which break glucose down into carbon dioxide and water; at the same time ATP is formed from ADP
- photosynthesis – enzymes are present which build up glucose molecules from carbon dioxide and water; at the same time they break down ATP into ADP to release energy
- protein synthesis – enzymes are required to help form the peptide links which join amino acids together.

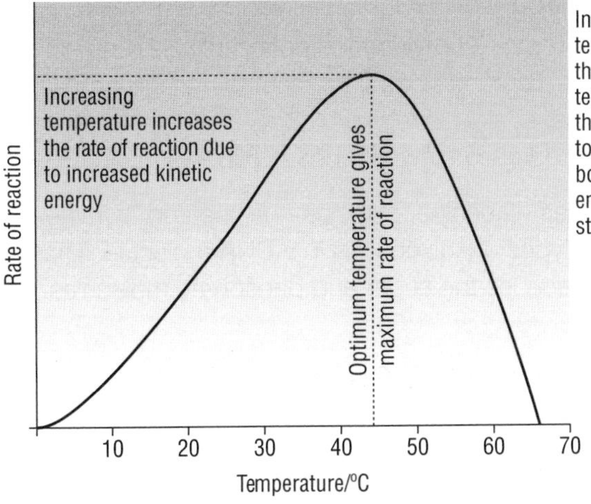

Figure 4 Graph to show effect of temperature on enzyme activity

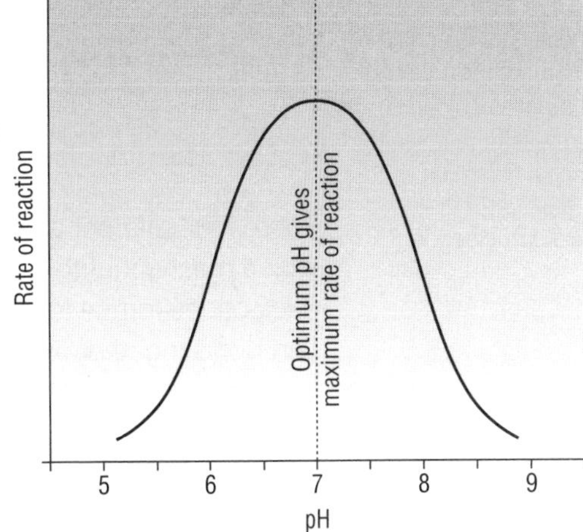

Figure 5 Graph to show effect of pH on enzyme activity

Questions

1 Explain the meaning of the following terms: substrate; active site; inhibitor.

2 The active site of an enzyme has a very precise three-dimensional shape. Explain why it is important to maintain this shape.

3 Enzyme activity is affected by the presence of inhibitors. State three other factors which affect the activity of all enzymes.

4 Sketch a graph (similar to Figures 4 and 5) to show how the activity of an enzyme depends on the concentration of the substrate.

2.3 (3) DNA and genes

Gene

A length of DNA which carries the codes to make a specific protein.

You have seen how the complex chemistry of every cell is made possible by a vast array of enzyme molecules, each responsible for catalysing a very specific reaction. But how does the cell synthesise these complex molecules accurately? The answer lies in the DNA molecules contained in the nucleus – they store the information for making proteins in structures called **genes**. This information can be passed on when cells divide.

DNA – deoxyribonucleic acid

DNA is found in the nucleus of cells – hence 'nucleic acid'. It is a long-chain molecule made up of many smaller sub-units called *nucleotides*.

The nucleotides themselves are made up of three components:

- a sugar molecule (deoxyribose)
- a base molecule, which contains several δ+ H atoms and δ– N and O atoms
- a phosphate group, which loses H^+ ions to become negatively charged – which explains why the DNA molecule is known as an acid.

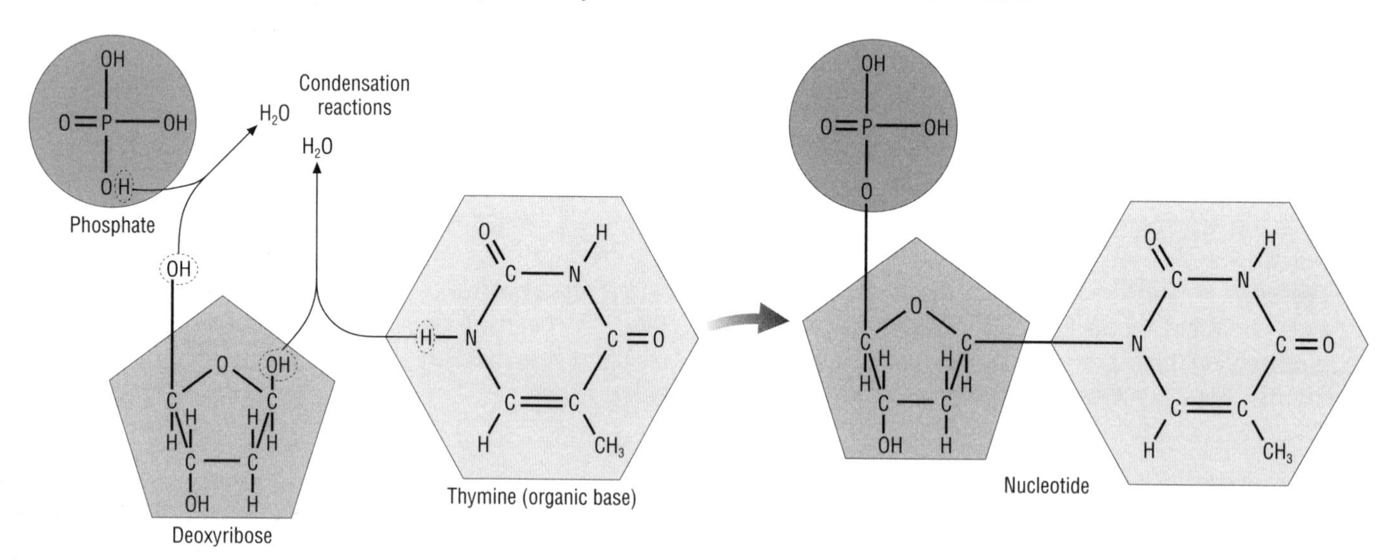

Figure 1 Phosphate, sugar and base unit join together to form a nucleotide

There four different types of base molecules – adenine, cytosine guanine and thymine – which means that there are four different types of nucleotides in a DNA molecule. These nucleotides form a long-chain molecule because the sugar and phosphate molecules can bond together.

The bases are often referred to by the initial letters A, C, G and T.

The structure of DNA

The discovery of the structure of DNA in 1953 was one of the most significant scientific breakthroughs of the twentieth century. This is because the way in which DNA carries and passes on information could be explained once the structure was known and understood.

DNA structure has several important features:

- there are *two* strands of nucleotides
- each strand consists of a *sugar–phosphate 'backbone'* with a base bonded to each sugar molecule
- the two strands are twisted round each other to form a *double helix*
- the strands are bonded together by the *bases* which *hydrogen bond* to each other
- the bases are found on the inside of the double-stranded molecule.

The base pairs

The most significant feature of the structure is that the bases bond to each other in a very specific pattern:

- adenine always bonds to thymine
- cytosine always bonds to guanine.

You can see from the diagrams that the atoms in these pairings are arranged in just the right way for hydrogen bonds to form. This is known as *complementary base-pairing*.

The other point to notice is that the two base pairs are almost exactly the same size and shape. This means that, whatever the order of bases in the DNA chain, the shape of the molecule remains regular with a uniform distance between the backbone of the two strands.

The A–T, C–G pattern means that if you know the sequence of bases on one strand it is possible to predict the complementary sequence on the other strand.

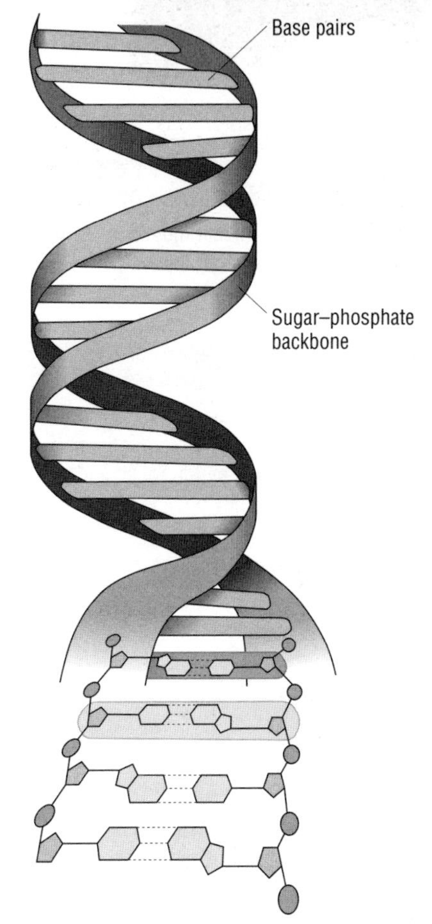

Figure 2 Structure of DNA

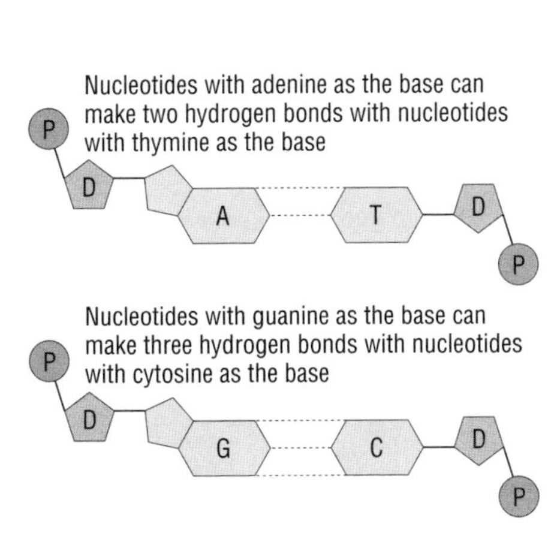

Figure 3 Hydrogen bonds in the base pairs

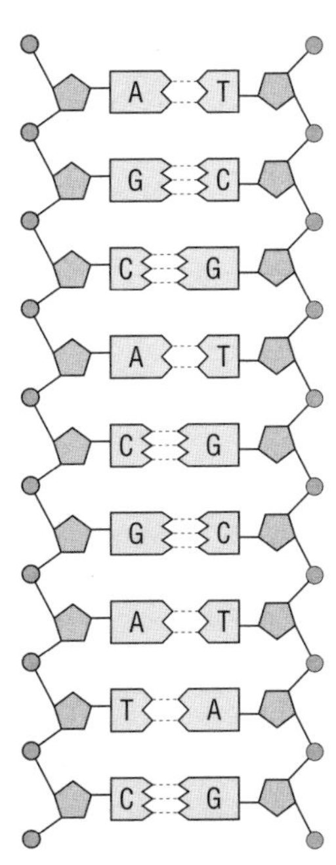

Figure 4 Sequence of base pairs on two strands of DNA

Examiner tip

You will not need to recall the structures of the bases, but you may need to use your understanding of hydrogen bonding to explain how they bond together.

Questions

1 DNA is a molecule which stores information. What is the information used for in a cell?

2 Representing a phosphate by P, bases by B and sugars by S, draw a diagram:
   (a) of a single nucleotide
   (b) to show how two strands, each consisting of three nucleotides, can hydrogen-bond together to form a short length of DNA.

3 The sequence of bases on one strand of DNA is ATTCGCCTTATG. Write out the sequence on the complementary strand.

4 Because DNA carries the information needed to make proteins, it is very important that the information does not get corrupted in any way. Suggest reasons why:
   (a) most DNA is found in the nucleus of cells, rather than in the cell cytoplasm
   (b) the bases, which carry the information, are on the inside of the double-stranded structure.

2.3 (4) DNA replication and transcription

The complementary base-pairing explained in spread 2.3.3 explains how DNA can copy itself when cells divide – this is known as DNA replication.

Replication

The copying of a single molecule of DNA to form two identical molecules.

Replication

When cells divide, the DNA in the original cell must be copied so that *identical* copies of the DNA can be passed on to the new daughter cells that are formed. This process of **replication** takes place in several stages:

- Enzymes cause hydrogen bonds between the base pairs to break, and the double helix begins to unwind. The two strands start to separate from each other and become exposed.
- New nucleotides from the surrounding environment form new complementary base pairs with the bases on the two exposed strands.
- Enzymes cause the sugar and phosphate groups of adjacent nucleotides to join together, creating a second new strand on each of the original strands.
- The process continues until there are two identical copies of the original DNA molecule.

Examiner tip

Many websites and textbooks give a more detailed description of replication than this. You will need to recall and describe only the key points as described here.

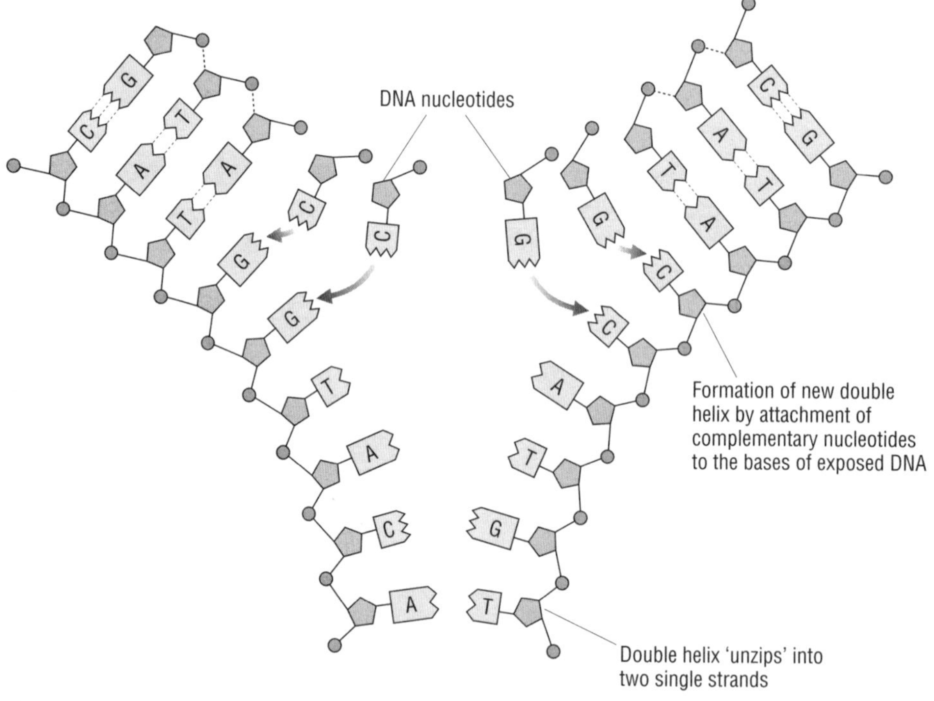

Figure 1 Replication

Activity

There are many excellent animations of DNA replication to be found on the Internet. Search using 'dna replication + animation'. These sites often also have animations of transcription and protein synthesis (see spread 2.3.5).

Protein synthesis

The ability to make identical copies of DNA explains one of the key functions of DNA – that it can be passed on when cells divide or organisms reproduce. The other key function is, of course, that DNA stores codes which act as a 'blueprint' for making specific proteins. The process of protein synthesis is more complex than replication but also makes use of the fact that complementary base-pairing allows the copying of information.

Ribosomes and messenger RNA

In eukaryotic cells, the DNA is mostly found in the nucleus. However, the synthesis of proteins occurs outside the nucleus at the ribosomes which are found on a region of folded membrane known as the *endoplasmic reticulum*. There must be a way of transporting information from the DNA in the nucleus to the ribosomes – that is the role of a molecule known as *messenger RNA* (mRNA).

RNA (ribonucleic acid) is another nucleic acid made up of a chain of nucleotides. There are some key differences between RNA and DNA:

- RNA contains the base uracil (U) in place of the thymine (T) present in DNA.
- RNA contains the sugar ribose in place of the deoxyribose in DNA.
- RNA is single-stranded whereas DNA is double-stranded.
- RNA is chemically less stable than DNA, so it is quickly broken back down into nucleotides after it has transferred the information to the ribosomes.

What is the information transferred by mRNA?

As you may have guessed, the key information necessary for protein synthesis is the sequence of bases in one of the DNA strands. mRNA has a complementary sequence of bases to this strand, and so it can be used to transport the information. The formation of mRNA takes place in a series of steps with some similarities to the replication process described above – it is called *transcription*:

- a small section of the DNA molecule unwinds
- new RNA nucleotides form complementary base pairs with one of the exposed DNA strands
- as the new RNA molecule begins to form it detaches itself from the DNA strand
- the two DNA strands rejoin.

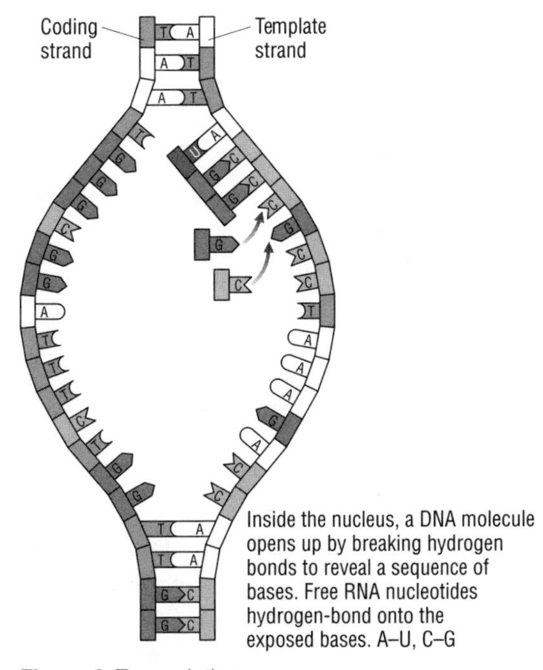

Figure 2 Transcription

Questions

1. Draw a table to compare DNA and RNA. Allocate the following features to DNA or RNA as appropriate – single-stranded; contains A, T, C, G bases; chemically very stable; contains A, U, C, G bases; contains ribose sugar; breaks down into nucleotides; contains deoxyribose sugars; double-stranded.
2. A short length of DNA has the base sequence TTACGCTATCCG. Predict the sequence of bases in the complementary mRNA strand formed from this DNA strand during transcription.
3. Explain the difference between the processes of transcription and replication. Under what circumstances does replication occur?
4. Give two ways in which enzymes are involved during the process of replication.

2.3 (5) The genetic code and protein synthesis

The genetic code

The information in DNA must code for the primary structure of a protein – that is, the sequence of amino acids. But DNA consists of a sequence of bases and there are only four bases. Proteins consist of a sequence of amino acids and there are 20 different amino acids found in the proteins of most living organisms. In fact, the information in DNA is stored not as a series of individual bases but as a sequence of three-base units or **triplet codons**. Each three-base unit carries the instruction to assemble a particular amino acid in the protein. For example UCU followed by GUC carries the instruction to assemble a serine amino acid followed by a valine amino acid.

Codon

A sequence of three bases in DNA or mRNA which specifies a particular amino acid in a protein.

So the sequence of codons in DNA determines the sequence of codons in mRNA, which determines the sequence of amino acids in a protein.

Figure 1 Information flow between codons and an amino acid sequence

Table 1 shows part of the genetic code. From this you can see which codons on mRNA specify particular amino acids in the final protein. In all there are 64 different codons, which code for the 20 different amino acids.

Codon	UCU	GUC	GCU	GGC	AAG
Amino acid	Serine	Valine	Alanine	Glycine	Lysine

Table 1 Some of the codons in the genetic code

Examiner tip

You may be asked to predict the sequence of codons in a mRNA molecule formed from a length of DNA, and then to predict the amino acid sequence in the protein which will eventually be formed. Information about the genetic code will always be provided if you need it.

The mechanism of protein synthesis

Although you can now see how the base sequence in DNA can act as a store of information that can eventually cause a particular amino acid sequence to be produced, the actual mechanism is rather more complex. It relies on one more type of nucleic acid – *transfer RNA* (tRNA). tRNA molecules act as a means of reading, or translating, the genetic code and they provide a way of bringing the amino acid molecules together in the correct sequence.

tRNA

tRNA molecules have a specific shape which allows them to fit into specific clefts or binding sites (rather like active sites) in ribosomes. They have two other key features:

- A three-base sequence, called an anticodon, at one end of the molecule.
- A site at the other end of the molecule to which an amino acid can be attached. The amino acid which attaches depends on the anticodon – for example a tRNA with an anticodon of UUU will only ever bond to the amino acid lysine.

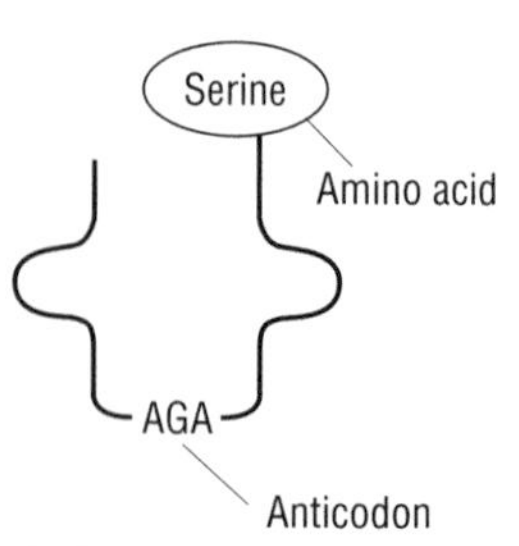

Figure 2 tRNA molecule showing anticodon (AGA) and amino acid attached (serine)

Translation

This is the process by which a sequence of codons on the mRNA directs the formation of a specific protein in the ribosome:

- An mRNA molecule attaches to a ribosome – two codons will fit into the binding sites in the ribosome.
- tRNA molecules bring amino acids to the ribosome. tRNA molecules with the correct, complementary anticodon will attach to the first codon of the mRNA.
- A second tRNA molecule brings the next amino acid to the second space in the binding site.
- A new peptide bond forms between the two amino acids.
- The bonded amino acids start to detach from the tRNA molecules, and the ribosome moves along the mRNA molecule, allowing new tRNA molecules to bind to the codons.

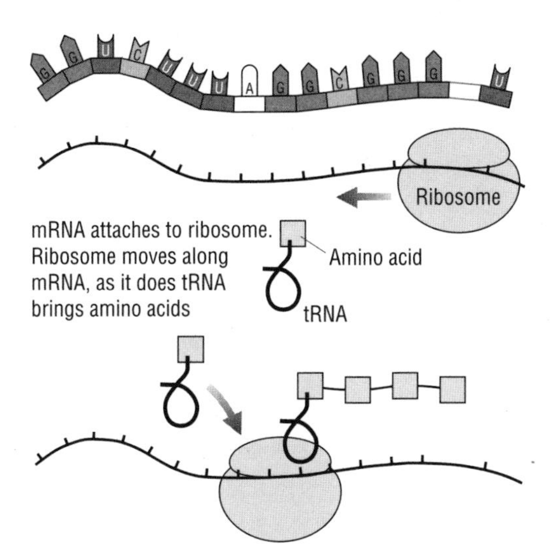

Figure 3 Translation

Questions

1. Draw a table to compare mRNA and tRNA. Allocate the following features to mRNA or tRNA as appropriate – contains a sequence of codons; contains one anticodon; formed in the nucleus when required; always present in the cytoplasm around the ribosome; attaches to a specific amino acid; diffuses from the nucleus to a ribosome.
2. A small section of a mRNA strand has the sequence GGCAAGGCUUCU.
   (a) How many codons does this strand contain?
   (b) Write down the sequence of anticodons which will bind to these codons.
   (c) Write down the sequence of amino acids in the protein which will be formed when the mRNA strand is translated.
3. DNA can sometimes become damaged in the nucleus. One of the ways in which it can be damaged is by two bases swapping places on a strand. This means that the mRNA formed from this DNA will also have bases in the wrong order – for example in the mRNA strand shown in question **2**, the seventh and eighth bases (G and the C) may end up in the wrong order. Explain the effect that will have on the resulting amino acid sequence in the final protein.
4. The genetic code is based on triplet codons – codons made up of sets of three bases. This means that there are 64 different codons possible. Suggest why it would not be possible for the genetic code to be based on codons consisting of just pairs of bases.

2.3 (6) Genetic engineering

How science works

In this spread you will consider examples of the possible benefits of genetically modified crops (HSW 6a).

Genome

A description of the sequence of bases present in an organism. This includes the genes, as well as long sequences of so-called 'junk' DNA that does not code for specific proteins.

Vector

A vehicle for introducing genes into a target organism.

Scientists now understand the link between the **genome** of an organism and the proteins that are produced using the instructions held in the base sequence. In turn, these proteins can affect the characteristics (observable features) of an organism (often called the *phenotype*).

In some cases, the connection between the genome, the protein and the characteristic is clear. For example, in the case of certain diseases we can identify the section of DNA in some human beings which codes for abnormal proteins. These abnormal proteins are responsible for some specific diseases, such as sickle-cell anaemia and cystic fibrosis. In other cases, it is not so easy to see the connection between gene, protein and characteristic. In any case, the environment to which the organism is exposed seems to affect whether or not the gene is 'expressed', and hence whether the characteristic appears in the organism.

Despite this, it is now possible to try to 'design' certain desirable characteristics into organisms by altering or engineering the genotype of the organism.

Genetically modified crops

The most controversial and environmentally significant application of this technique has been the use of genetic engineering to produce *genetically modified* (GM) crops. In this process, genes from a completely unrelated species (the donor organism) are transferred into the genome of a crop plant in order to improve the characteristics of the crop plant. To do this, certain things must happen:

- the gene for the required characteristic must be identified and extracted from the donor organism
- this gene must be inserted into the cells of the target organism (the crop plant)
- the cells which carry this gene must be identified and grown on into complete plants.

The easiest characteristic to transfer is one that relies on the presence of a single protein – for example the ability to repel an insect pest might be due to the presence of a toxic protein, or resistance to a herbicide might be due to the presence of an enzyme which causes the breakdown of the herbicide molecule.

How genetic modification is carried out

A gene can be cut out of the genome of the donor organism by using *restriction enzymes*, which cut DNA at very specific base sequences.

This DNA must then be transferred into the genome of the target cells. There are three possible ways of doing this, using different vectors:

- Inserting the gene into a plasmid – a small circular piece of DNA – taken from a bacterium, such as *Agrobacterium tumefaciens*, which infects plant cells and in doing so incorporates its genes into the plant genome.
- Inserting the gene into a virus – viruses infect plant cells in a similar way.
- Attaching the gene to minute particles of an inert substance, such as gold, and injecting these into the nucleus of the target cells.

It can be difficult to be certain whether the required gene has been successfully transferred or not. One possibility is to grow the cell into a full-sized plant and see if it shows the desired characteristic. But this would be expensive and time consuming. Marker genes are used instead – these are genes which are transferred at the same time as the desired gene. The marker gene might be a gene which causes the cells to fluoresce under ultraviolet light or for antibiotic resistance – in which case the GM cells can be identified because they will not be killed when they are exposed to the antibiotic. Once the GM cells have been identified, techniques such as tissue culture (growing the cells on a special medium) are used to allow the cells to develop into fully grown plants.

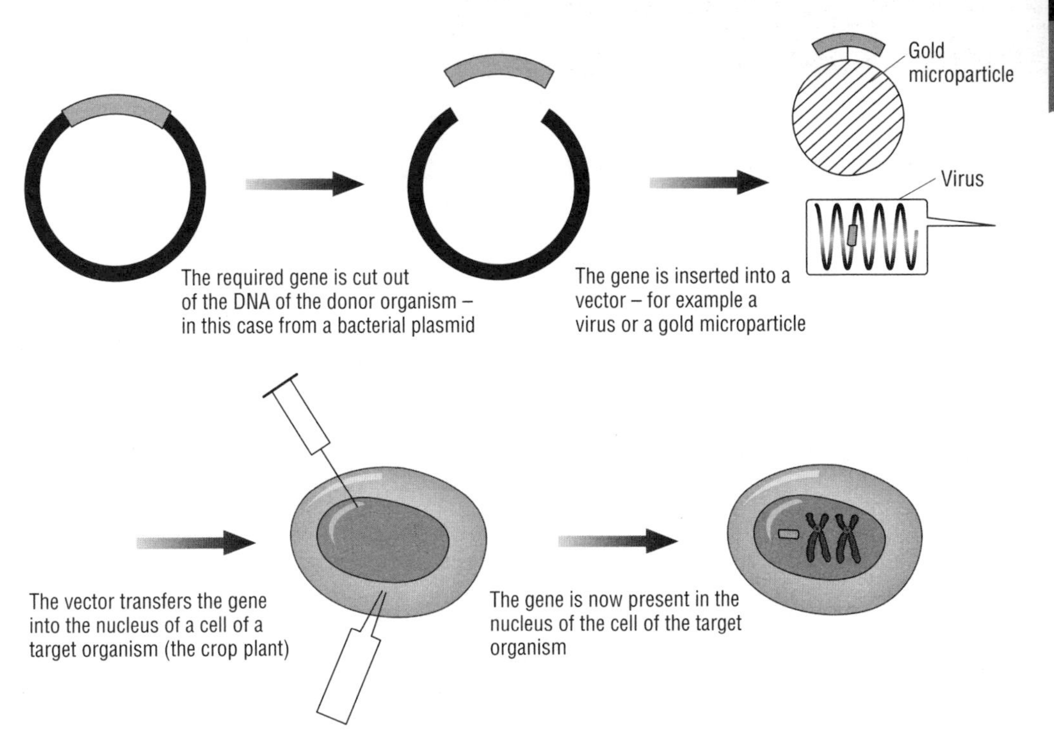

Figure 1 Stages in genetic engineering

Applications: Roundup resistance

Roundup is one of the most commonly used herbicides (weedkillers) in the world. However, it can kill some crop plants as well as weeds, so it cannot be used when crop plants are actively growing.

A gene from a bacterium is introduced into some plants such as cotton or tobacco – this gene seems to provide resistance to the herbicide. As a result, *Roundup* can be applied while these GM crops are growing.

Applications: Bt corn – introduced pest resistance

Maize plants can be damaged badly by caterpillars which bore into the leaves and stems of the plants. The caterpillars can be killed by spraying with *Bacillus thuringiensis*, which produces a protein (the Bt toxin) toxic to the caterpillar. By introducing the gene for the Bt protein into the maize plant, the cells of the maize plants themselves become toxic to the caterpillar. Bt toxins only affect insect larvae, not vertebrates.

Applications: golden rice – a more nutritious crop

Genes derived from daffodils have been introduced into rice. The genes code for the enzymes required in the pathway that cells use to synthesise β-carotene. This is converted into vitamin A by the human body, so this rice could help to reduce vitamin A deficiency in third world countries.

Activity

There is a huge range of material available on the Internet providing information about other uses of GM crops, and also describing the environmental concerns. Try to find three other examples of the successful use of GM crops – and also three different pieces of evidence used in opposition to the use of GM crops.

Questions

1 (a) Explain the meaning of the word 'vector' as used in genetic engineering.
   (b) Give three vectors used in genetic engineering.
2 (a) What is a 'marker gene' and why are they used?
   (b) Give an example of a marker gene which could be used in genetic modification of plants.
3 It is normally much easier to introduce a gene into a bacterium than into a eukaryotic cell such as a plant cell. Suggest why this is.
4 State three ways in which crop plants have been genetically modified. In each case state why the modification is an advantage.

2.3 (7) Concerns about genetically modified crops

How science works

In this spread you will consider the social, ethical and environmental implications of the use of genetically modified crops (HSW 6b).

You saw in spread 2.3.6 how genetic modification has been used to introduce some specific advantages into crop plants. The technology to do this on a large scale has been available since the 1990s, but in many countries there remains controversy about whether GM crops should be introduced on a commercial scale.

The arguments for the use of GM crops

Many organisations and individuals feel that the benefits of GM crops to farmers and consumers are considerable. In particular, they often make the following points:

- Characteristics such as resistance to insect pests allow higher and more reliable yields to be obtained from crops, bringing prices down for the consumer and making farming a more viable activity.
- GM can enable the quality of the crop to be increased – for example, increasing the nutrient levels in the crop, improving the storage properties or even enhancing the taste, colour or texture.
- In order to feed the increasing human population successfully, crops may need to be grown on marginal land or in climatic conditions that normally limit the yield of crops. It may be possible to make crops grow better in these conditions, reducing food shortages and famines.
- If GM crops can resist insect pests there might be a reduced need to use chemical insecticides. These insecticides are damaging to the environment and might even be toxic to human beings, so reducing their use would be a major advantage.
- Manipulating genes is not a new process – human beings have been altering the genome of their farmed animals and crops for thousands of years by selective breeding. And transgenic bacteria are routinely used to produce, for example, medical products such as human insulin.

Figure 1 Maize – a crop which has been genetically modified

The arguments against the use of GM crops

Opponents of GM crops use environmental arguments, as well as issues of consumer safety and ethics, to make their case:

- The genes transferred into GM crops may spread to other crop plants. It is possible for pollen from the flowers of GM plants to travel quite large distances on the wind. So genes from GM maize could spread into 'conventional' maize being grown by a farmer several miles away. Many farmers object to this – particularly if they have 'organic' status.

- Of even more concern is the possibility that these genes could spread into the wild plant community. Some wild plants are closely related to crop plants – for example wild mustard is related to oil seed rape – so it is possible for GM crops to cross with these wild plants. If this happened the resulting plant might possess the same characteristic as the GM plant – for example resistance to pests or the ability to grow in drought conditions. These plants are often described as 'superweeds' – wild plants which would grow very effectively in the environment and would be extremely difficult to eradicate – so the introduction of GM crops might actually increase the use of herbicide chemicals.
- The use of antibiotic-resistant marker genes concerns many people because bacteria are able to absorb genes from their environment. So, if we eat GM food which contains a gene for antibiotic resistance then bacteria in our gut may absorb this gene and become resistant to that antibiotic.
- The new gene in the organism may have other consequences for the metabolism of the plant. This might lead to the production of substances which are toxic or allergenic (cause an allergic reaction). Although there is no evidence of human health being affected by the GM crops produced so far, thorough testing would be required to be certain that the food is safe.
- Many people suspect that the benefits to farmers and consumers may not be as great as the biotech companies claim. GM crops are patented, so individual companies have a monopoly allowing them to charge whatever price they choose. The extra price benefit to farmers and consumers of higher yielding crops may be outweighed by the increased cost to the farmer of buying GM seed. And in developing countries the price of the new seed may exclude many of the poorer farmers who may need the product most.

What do you think?

This topic provides an excellent opportunity for you to use your research skills to find out more about the arguments. Which arguments do you find the most powerful? How would you respond to some of the arguments laid out here? You may be given the chance to take part in a formal debate on the issue.

Figure 2 There are environmental concerns about growing GM crops

Activity

You may choose to set up a role-play debate to feedback your findings about GM crops. Four students can volunteer to play specific roles in the debate representing, for example:

- a scientist from a biotechnology company seeking permission to market a GM crop
- a representative of a consumer company worried about food safety
- a representative of a development charity considering the impact of GM on farmers in the developing world
- a member of an environmental group concerned about environmental issues in the UK if GM crops are grown.

Each student should research the arguments likely to be used by each of these people and present them in a debate. Other students can research possible questions to put at the end of the presentations.

Question

1 Write an essay in which you explain the issues connected with the GM crop debate. In the essay you should:
  - **(a)** explain what is meant by a GM crop
  - **(b)** give some specific examples and explain the benefits they may have
  - **(c)** outline some arguments used against the introduction of GM crops.

2.3 Module 3 – Summary

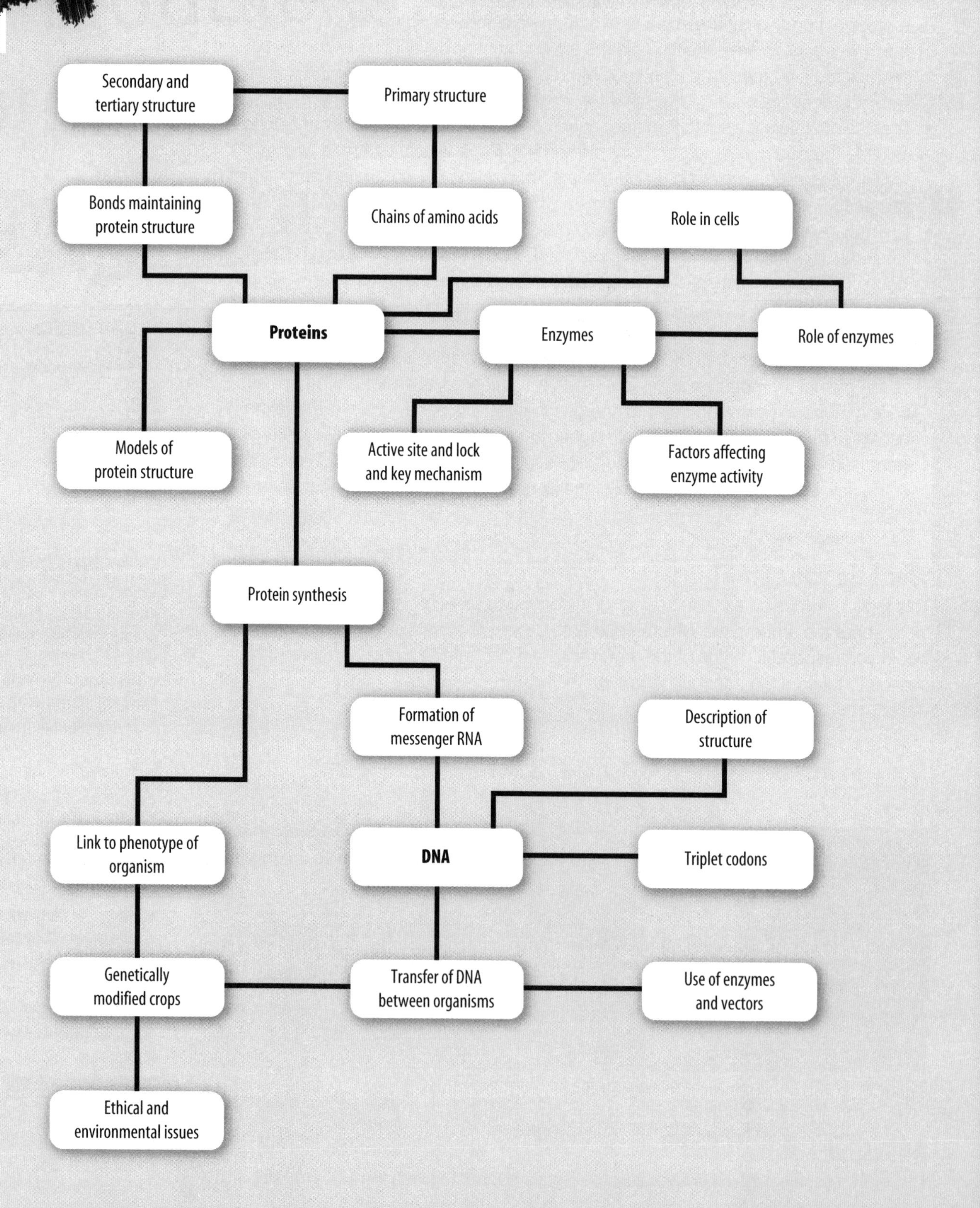

Practice questions

Low demand questions

These are the sort of questions that test your knowledge and understanding at E and E/U level.

1 One of the roles that proteins have in living organisms cells is that they act as enzymes.

(a) (i) Describe the importance of enzymes to living organisms.

(ii) State two other roles of proteins in living organisms.

(b) Figure 1 is a diagram of a protein – it shows the active site of the enzyme.

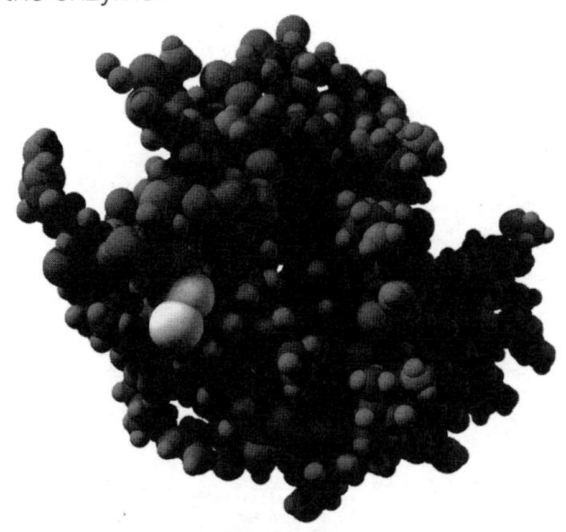

Figure 1 Space-filling model of an enzyme

(i) Explain why the presence of an active site is important in the action of an enzyme.

(ii) Describe two ways in which the active site could alter in shape.

(iii) What is the effect on enzyme activity of a change in the shape of the active site?

2 Genetic modification of crops is one way in which it is hoped to allow more food to be grown in harsh environments.

(a) Describe two ways in which crops might be altered using genetic engineering which could allow more reliable food production.

(b) In order to carry out the process of genetic modification, a gene must be introduced into the genome of a crop plant using a vector. Give two possible vectors which could be used.

Moderate demand questions

These are the sort of questions that test your knowledge and understanding at C/D level.

3 The sequence of bases in one strand of DNA is CGCTTACGG.

(a) (i) Write down the sequence of bases in the complementary strand of this DNA strand.

(ii) How many codons are contained in this sequence of bases?

(iii) What is the significance for protein synthesis of the sequence of bases in a codon?

4 Figure 2 shows a ribbon diagram of an enzyme.

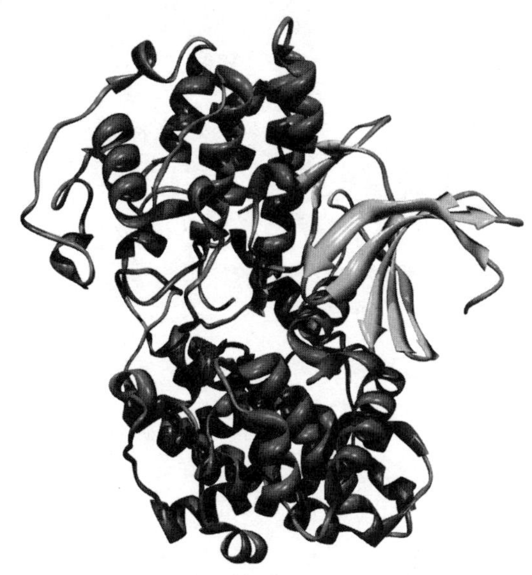

Figure 2 A ribbon diagram of an enzyme

(a) The primary structure of an enzyme is the sequence of amino acids present. Use Figure 2 to explain what is meant by:

(i) the secondary structure of the enzyme

(ii) the tertiary structure of the enzyme.

(b) The tertiary structure is maintained by a variety of forces including hydrogen bonds, covalent bonds and ionic bonds. What types of forces are responsible for maintaining:

(i) the primary structure

(ii) the secondary structure?

High demand questions

This is the sort of question that tests your knowledge and understanding at A/B level.

5 In genetic engineering, genes are transferred from the genome of one organism to the genome of another, different, organism. Sometimes a 'marker' gene is also transferred at the same time. This makes it possible to check if gene transfer has been successful.

(a) What is meant by the term 'genome'?

(b) Give an example of the use of a marker gene in genetic engineering, and describe how it would be used to check whether gene transfer has been successful.

(c) Some people are concerned that the transfer of the marker gene could cause environmental problems. However, there are other possible environmental problems which could potentially result from the increased use of genetically modified crops. Describe one such problem.

UNIT 2

Module 4 Options for energy generation

Introduction

In this module you will look at some of big questions which mankind must ask itself about the way in which the Earth's resources can be used sustainably in order to generate usable energy. In modern society that effectively means electrical energy.

You will first see why it is that the fossil fuels – coal, natural gas and oil – are such an effective store of energy by finding out more about the way that chemical bonds are broken and formed when they burn. You can then compare this with a newer technology for energy generation – the use of nuclear fission. This will lead you into the story of how the nucleus was first discovered and how our model of the atom changed over time. The nuclear model of the atom allowed scientists to make sense of the mysterious phenomenon of radioactivity, and you will use ideas about radioactivity to assess the safety of nuclear power as an alternative to fossil-fuel burning.

We then turn our attention to the way in which the generated electrical energy is transmitted to the consumers of the energy. This will allow you to remind yourself about some basic ideas about electrical circuits from GCSE, but also to understand how the concept of electromagnetic fields means that the processes that go on in an electrical circuit may affect a much wider region.

The potential health risks from electromagnetic fields are considered and you will find out how scientists attempt to assess those risks.

How science works

During this module you will cover some of the aspects of How Science Works. In particular, you will be studying material which may be assessed for:

- HSW 5a: Carry out experimental and investigative activities in a range of contexts.
- HSW 5b: Analyse and interpret data to provide evidence, recognising correlations and causal relationships.
- HSW 5c: Evaluate methodology, evidence and data and resolve conflicting evidence.
- HSW 6a: Consider applications and implications of science and appreciate their associated benefits and risks.
- HSW 7a: Appreciate the tentative nature of scientific knowledge.
- HSW 7b: Appreciate the role of the scientific community in validating new knowledge and ensuring integrity.
- HSW 7c: Appreciate the ways in which society uses science to inform decision making.

Examples of this include:

- Obtain and interpret results from experiments to measure the energy released by burning fuels (spread 2.4.1).
- Discuss how scientists use models of systems that cannot be investigated directly (spreads 2.4.3 and 2.4.4).
- Evaluate nuclear fusion *including cold fusion* (as an option for future energy generation (spread 2.4.8).
- Describe the options for future energy generation strategies and assess their relative advantages and disadvantages (spreads 2.4.2, 2.4.7 and 2.4.8).
- Describe and compare the hazards associated with radioactive sources (spreads 2.4.5 and 2.4.6).
- Describe the use of epidemiological studies to assess the risks of exposure to alternating electromagnetic fields (spreads 2.4.12 and 2.4.13).

Module contents

1. **Burning fuels**
2. **Fossil fuels**
3. **The story of the atom**
4. **The nucleus**
5. **Radioactivity**
6. **Nuclear processes**
7. **Options for future energy generation (I)**
8. **Options for future energy generation (II)**
9. **Electricity transmission and distribution**
10. **Electrical circuits**
11. **Electrical and magnetic fields**
12. **Alternating fields and epidemiology**
13. **Analysing epidemiological data**

Test yourself

1. How are fossil fuels, such as oil and coal, formed?
2. What are the products when a fuel, such as methane, is burned?
3. What sub-atomic particles make up atoms? Where in the atom are they found?
4. Name the three types of radiation which can be emitted by a radioactive substance.
5. What is an electric current? What name is given to materials which allow an electric current to flow easily?
6. How are the terms 'power' and 'energy' related?

2.4 ① Burning fuels

How science works

In this spread you will be introduced to a simple experimental technique for measuring the energy released by burning fuels (HSW 5a).

Combustion

A reaction in which a fuel reacts with oxygen to produce heat energy – we often describe a combustion reaction as simply 'burning'.

Fossil fuel

A substance formed from the fossilised remains of dead plants and animals, which can take part in a combustion reaction to produce heat.

Modern society across the globe is reliant on the availability of energy, particularly in the form of electrical energy. Since the Industrial Revolution in the eighteenth century, the source of this energy has been the **combustion** of **fossil fuel** reserves which exist under the Earth's surface.

Bond breaking and forming

You saw earlier (spread 2.1.4) that a covalent bond involves the attraction between two positive nuclei and a pair of negative electrons. To separate a nucleus from the electrons, and break the bond, will require energy to be provided, for example in the form of heat. So, *bond breaking requires energy*.

That may seem fairly obvious, but it is perhaps less obvious that *bond forming releases energy*. When the positive nucleus and the electrons are far apart, they possess stored energy (known as potential energy). When they move together, this energy is transferred into movement energy and eventually becomes heat energy (thermal energy).

When a fossil fuel, such as methane, burns in air, both bond breaking and bond forming occur during the reaction.

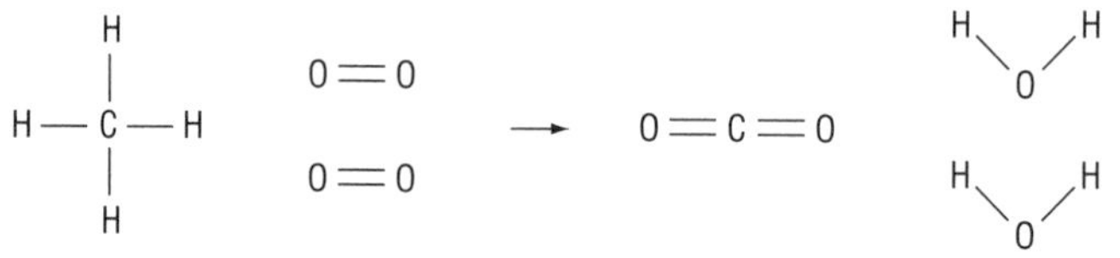

C — H and O = O bonds are broken; C = O and O — H bonds are formed

Figure 1 Combustion of methane showing structural formulae of reactants and products

Some bonds (e.g. C–H and O=O) are broken, and others (e.g. C=O and O–H) are formed. In this case, the energy released by forming bonds is greater than the energy required to break bonds, so the reaction releases energy (in the form of heat). This is known as an exothermic reaction. The release of heat in this way is why methane can be described as a fuel.

Measuring the energy released by fuels

You may have done experiments in which fuels are used to heat up water. This provides a simple way of measuring the energy released by different fuels.

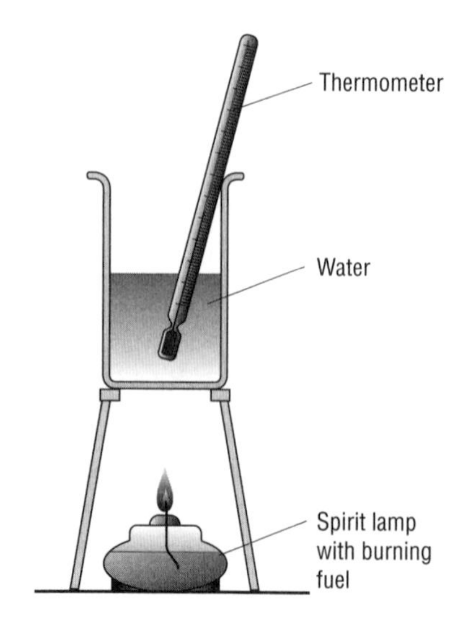

Figure 2 Simple calorimetry experiment to measure the heat released by burning a fuel

Activity

Liquid fuels can easily be burnt in small spirit lamps and used to heat up water in a metal can. Carry out experiments to compare the amount of energy released by different fuels.

How much energy is released by burning 1 kg of ethanol?

Ethanol is now often added to petrol because it is a renewable biofuel. In an experiment, 0.50 g of ethanol was burnt and the temperature of 100 g of water increased from 19 to 41 °C. Calculate how much heat energy, in kJ, could be released from 1 kg of ethanol.

The specific heat capacity of water is $4.2\ J°C^{-1}g^{-1}$.

The temperature rise of the water was 22 °C.

4.2 J is needed to raise the temperature of 1 g of water by 1 °C, so the energy needed to raise the temperature of 100 g of water by 22 °C is $4.2 \times 100 \times 22 = 9240$ J.

0.50 g of ethanol releases 9240 J of energy.

1000 g of ethanol releases $9240 \times 2000\ J = 1.848 \times 10^7\ J = 1.848 \times 10^4\ kJ$.

Can you think of any reasons why this experiment might tend to produce a *lower* number than the data book value for ethanol?

Examiner tip

Always set out your working in a logical way like this. That way, if you make a mistake early on the examiner can see that you can still do the rest of the calculation accurately.

Comparing fuels

The figure we calculated in the worked example is often called the 'energy density' – the energy released from 1 kg of fuel. Values of energy density can be compared for a range of fuels.

Fuel	Chemical formula	Energy density/$kJ\ kg^{-1}$
Hexane (similar to petrol)	C_6H_{14}	4.84×10^4
Methane (natural gas)	CH_4	5.56×10^4
Carbon (coal)	C	3.28×10^4
Hydrogen (a non-fossil fuel)	H_2	14.3×10^4

You will notice that many fossil fuels have quite similar values for energy density – you will find out more about the use of hydrogen as a fuel in spread 2.4.2.

Questions

1 Name three fossil fuels.

2 Some processes, such as the condensation of water, involve only the formation of bonds (in this case O–H bonds). Predict whether the condensation of water requires energy or releases energy.

3 (a) If 1 kg of carbon and 1 kg of methane were burnt as fuels, which one would release the most energy?

(b) Despite this, coal is often described as a more concentrated energy source than methane. Suggest why.

4 In an experiment to compare the energy density of some fuels, 1.0 g of methanol was burnt. The temperature of 200 g of water rose by 18.7 °C. Calculate a value for the energy density of methanol in $kJ\ kg^{-1}$. (Hint: the specific heat capacity of water is $4.2\ J°C^{-1}g^{-1}$.)

2.4 (2) Fossil fuels

The three main fossil fuels – coal, oil and natural gas – together provide approximately 80% of the world's energy demands. The remaining 20% is accounted for by nuclear energy and renewable sources – mostly hydroelectric.

How science works

In this spread you will consider the environmental consequences of burning fossil fuels (HSW 6a).

Activity

Find figures which give you a value for the total world energy demand. Searching for images of 'world energy demand' will give you graphs and pie charts to analyse.

Look at the trends in total energy demand over the past 50 years and compare how different sources of energy are used to meet that demand.

Examiner tip

You may be asked to analyse and present data in a number of different ways – for example to produce bar charts or pie charts from data, or to comment on the patterns shown in a given graph.

Figure 1 Cooling tower at a power station

Total fossil fuel use is estimated at about 8000×10^6 tonnes of 'oil equivalent'. In other words, the energy released from the burning of fossil fuels is equivalent to burning this mass of oil. In fact, the proportion of oil being burnt is gradually falling – partly because reserves of oil are becoming depleted or more difficult to access because of political issues or conflict.

Environmental consequences

Carbon dioxide emissions

All fossil fuels contain carbon and so the burning of all fossil fuels will release carbon dioxide – water vapour is also formed in most cases. The chemical equations for the combustion of fossil fuels can be compared below:

- Coal $C + O_2 \rightarrow CO_2$
- Natural gas $CH_4 + 2O_2 \rightarrow CO_2 + 2H_2O$
- Petrol $C_5H_{12} + 8O_2 \rightarrow 5CO_2 + 6H_2O$
- Fuel oil (used in oil-fired power stations) $C_{21}H_{44} + 32O_2 \rightarrow 21CO_2 + 22H_2O$

Hydrocarbon

A molecule consisting of carbon and hydrogen only.

In fact, many fuels are actually a mixture of hydrocarbons so these equations are only examples of some of the reactions which occur when the fuel is burnt.

Examiner tip

You may be asked to complete a balanced chemical equation (see spread 2.2.2) for the burning of fossil fuels, particularly hydrocarbons.

All of these reactions produce carbon dioxide. Notice that in the case of a hydrocarbon (such as natural gas and petrol) the hydrogen atoms in the fuel are oxidised to water. So overall slightly less CO_2 is released per tonne of hydrocarbon fuel than for coal. You have learnt about the way in which CO_2 can cause warming of the atmosphere in several earlier spreads.

Balancing equations for the combustion of hydrocarbons

Complete this equation for the combustion of a molecule of heptane (C_7H_{16}) which is a component of petrol:

C_7H_{16} + ___O_2 → ___CO_2 + ___H_2O

Follow these steps to balance the equation:

- Balance the C atoms – there are 7 C atoms in the heptane molecule, so there must be 7 C atoms on the right-hand side. So there will be 7 CO_2 molecules.
- Balance the H atoms – there are 16 H atoms on the left-hand side, so there must be 16 H atoms on the right-hand side. So there will be 8 H_2O molecules.
- Finally deal with the oxygen – $7CO_2$ contain 14 O atoms and $8H_2O$ a further 8 O atoms, so that's 22 O atoms in all. So there must be 11 O_2 molecules on the left-hand side.

So the balanced equations is

$$C_7H_{16} + 11O_2 \rightarrow 7CO_2 + 8H_2O$$

Remember the order for balancing: C → H → O.

Figure 2 Oil platform in the North Sea

SO_x and NO_x

Fossil fuels are formed from the remains of animals and plants, chemically changed by the effects of heat and high pressure over millions of years. Because biomass, particularly protein biomass, also contains sulfur and nitrogen atoms, many fossil fuels also contain these atoms chemically combined with hydrogen and carbon atoms. When the fossil fuel is burnt, the sulfur and nitrogen atoms are oxidised to SO_x and NO_x. You learnt about how these gases can cause acid deposition in spread 2.2.2.

Coal often contains a high level of these sulfur impurities. There is particular concern about the amount of high-sulfur coal which may be burnt in China to meet that country's expanding energy needs.

Nitrogen oxides are also involved in the formation of photochemical smog in the atmosphere. The nitrogen oxides react with other pollutants, such as hydrocarbons, to produce a toxic mixture of gases and particulates which cause breathing problems and asthma.

Other effects

Fossil fuels have to be extracted from underground reserves. The mining of these reserves and the transportation by road, rail, sea or pipeline all increase the environmental burden – as does the highly energy-intensive way in which it is processed to generate the fuels that we use.

Figure 3 Oil drums

Questions

1. All fossil fuels produce CO_2 when they burn. Explain why some fuels produce less CO_2 (per tonne of fuel burnt) than others.
2. What is meant by the term 'high-sulfur coal'? Why is the burning of high-sulfur coal an environmental concern?
3. Propane is a fuel used in camping gas stoves. Write a balanced chemical equation to show the reaction of propane (C_3H_8) with oxygen to produce carbon dioxide and water.

2.4 ③ The story of the atom

How science works

In this spread you will discover how successive models of the atom have been accepted by the scientific community (HSW 7b).

Activity

You can follow the story of the development of the model of the atom on several websites. Start off the story by searching for 'Cambridge physics plum pudding'.

Applications: the development of models

In spread 2.1.2 you came across the molecular kinetic model of a gas, which helped scientists to visualise and explain the behaviour of gases. In this model, all the particles, including atoms, are treated as though they are simple, hard spheres that can collide with each other but never change. This is often described as the 'billiard ball' model of the atom. This works well in explaining the behaviour of gases but it does not explain many other properties of atoms. The model used to explain these properties developed rapidly at the beginning of the twentieth century as new discoveries were made. The atoms could never be observed directly – instead scientists improved the models as new discoveries and observations were made.

Fossil fuels are described as a *finite resource* – there is a limited amount of fossil fuels stored under the surface of the Earth. The rate at which new fossil fuels can be formed is much slower than the rate at which they are being depleted, so reserves will run out at some point in future, or the price will rise so much that it is no longer economic to generate energy from them. This could begin happen as early as the end of this century.

Other, very different, methods have been proposed to generate energy. Two examples are nuclear fission, which has been used for several decades by some countries, and nuclear fusion, which will take many more years of research before we know if it is a viable source of energy.

Both of these make use of processes which affect the very structure of the atom. The story of the way in which this structure was discovered is a very important one in the history of science.

The Thompson plum pudding model

When the electron was discovered in 1897 by J.J. Thompson, a British scientist, it was realised that atoms must have an internal structure – if they contained negative particles (electrons) there must by something positive to balance out the charge. Thompson suggested that the negative electrons were embedded in a sphere of positive matter – rather like the plums in a fruit pudding. This model replaced the simple 'billiard ball' model of the atom.

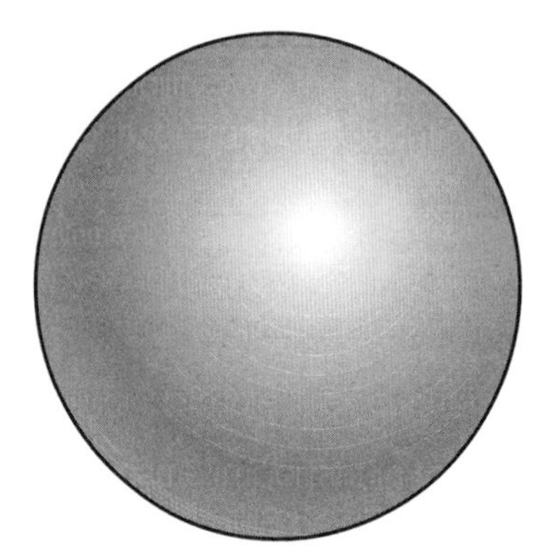

Figure 1 Billiard ball model of the atom – the atom as a spherical particle that cannot be broken down any further

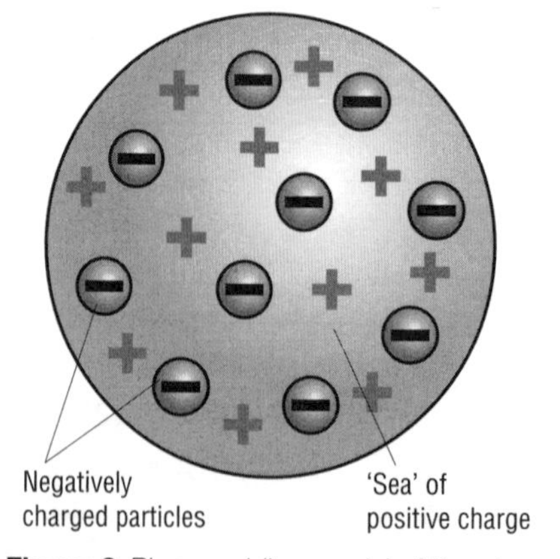

Figure 2 Plum pudding model of the atom

The discovery of radioactivity and alpha particle scattering

Radioactivity plays an important part later in the story, but it was the discovery that some atoms produce particles called alpha particles which led to the next crucial observation.

Ernest Rutherford, a scientist born in New Zealand but working in England, found that the alpha particles emitted by some radioactive substances could pass through thin films of metal or mica (a type of ceramic). However, the metal seemed to have a strange effect on the alpha particles – a bit like the scattering of light.

Two of Rutherford's researchers, Hans Geiger and Ernest Marsden, found a much more astonishing effect. When they fired alpha particles at a very thin sheet of gold leaf, some of the alpha particles were scattered through very large angles, even bouncing back towards the source.

Radioactivity

The ability of some types of unstable atom to emit energy in the form of particles or high frequency electromagnetic waves.

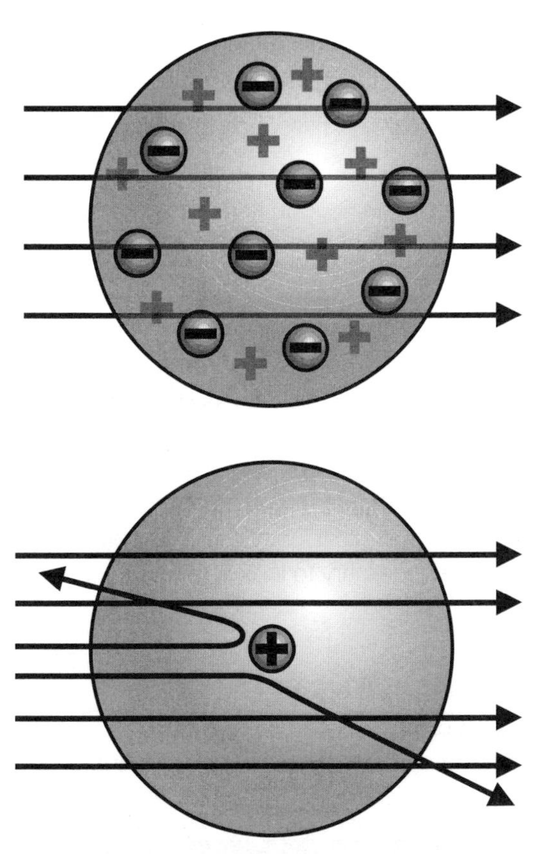

Figure 3 The scattering of alpha particles could only be explained by the model in which the positive charge was concentrated in a nucleus

Rutherford knew that alpha particles were positively charged. The only way to explain how this scattering occurred was to suppose that all the positive charge in the gold atoms was concentrated in a very small space. Then the repulsion between the two positively charged objects would be strong enough to cause the alpha particles to 'bounce' back. He called this small positive region the *nucleus* and proposed that this must be at the centre of the atom, with the electrons surrounding it. This was known as the *nuclear model.*

Examiner tip

The word 'nucleus' can be confusing because it is used to describe both the positive centre of an atom and the organelle in a cell which controls its behaviour. Hopefully it should be fairly clear from the context which type of nucleus is being discussed! It is also a little awkward to find a suitable word to use for the plural of nucleus – scientists use 'nuclei'.

Question

1 Put these observations, or models, into chronological order starting with the earliest. Then match each up with the scientist (or scientists) responsible – alpha particles scattering through large angles; the nuclear model; discovery of the electron; discovery that alpha particles could pass through matter; the plum pudding model.

2.4 ④ The nucleus

How science works

In this spread you will consider the validity of different models of atoms used by scientists (HSW 7a).

The model which Rutherford proposed for the atom – a nucleus surrounded by electrons – was not the end of the story.

Rutherford himself discovered that hydrogen nuclei were sometimes produced when alpha particles collided with nitrogen atoms. He called the hydrogen nucleus a *proton* and suggested that protons are present in all nuclei. It is the protons which give the nucleus its positive charge.

He went on to suggest that, because the mass of most nuclei was greater than the mass of the protons they contained, there must be another particle present, which he called the *neutron* (because it was uncharged). This particle was not actually detected until 1930.

Niels Bohr, a Danish physicist, suggested that the electrons would surround the nucleus in fixed orbits, or shells. This has proved very helpful to chemists in explaining the chemical properties of atoms.

Physicists now know that the simple neutron and proton particles suggested by Rutherford are themselves made up of even simpler particles known as quarks. Many other sub-atomic particles also exist but are only detectable under extreme conditions.

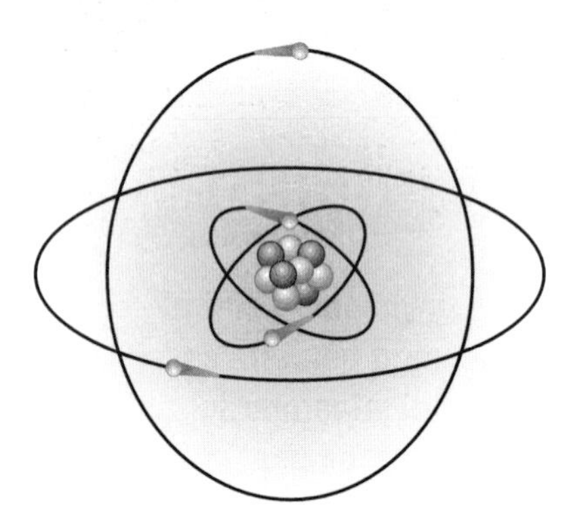

Figure 1 Rutherford–Bohr model: protons and neutrons in the nucleus and electrons in fixed shells

Which one is right?

This is a difficult question to answer. In science it is sometimes impossible to say that a theory is 'right' or 'wrong' in all situations. In different situations, we may choose to use different models – the pressure of a gas can be explained very well by using the simple billiard ball model; the Bohr atom may be more helpful in explaining the chemical behaviour of atoms. Only in the extreme conditions of experiments carried out by some nuclear physicists would it be necessary to look at the quark structure of the particles in the nucleus.

The properties of the particles in the atom

The masses and charges of protons, neutrons and electrons are compared in Table 1. To make it easier, all the data are compared to the mass and charge of a hydrogen nucleus (mass = 1.67×10^{-27} kg; charge = 1.60×10^{-19} C).

	Mass ($^1H = 1$)	Charge ($^1H = +1$)
Proton	1	+1
Neutron	1	0
Electron	0.00055	−1

Table 1

Atomic number

The number of protons in a nucleus – sometimes called the proton number.

Mass number

The total number of protons and neutrons in a nucleus – sometimes called the nucleon number.

Describing atoms

Scientists use the terms **atomic number** and **mass number** to describe the nucleus of an atom. These can be shown in a nuclear symbol.

For example, the most common form of a phosphorus atom is represented as $^{31}_{15}P$.

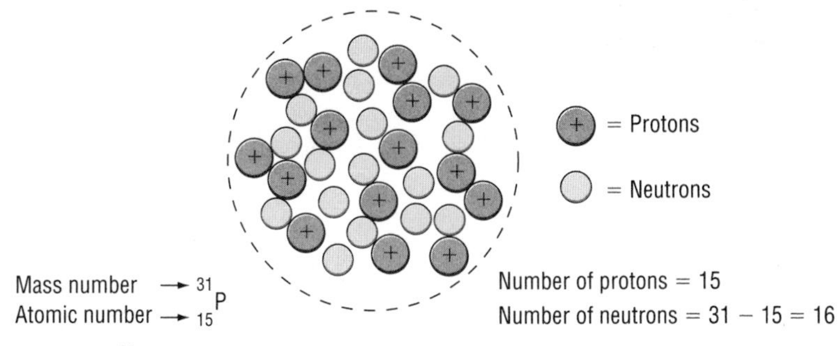

Figure 2 $^{31}_{15}P$ nucleus

Isotopes

All atoms of the same element have the same number of protons – so, for example, all carbon atoms have 6 protons. But the number of neutrons can differ – most carbon atoms have 6 neutrons, but some have 7 and others have 8 neutrons. These atoms are called **isotopes** of carbon.

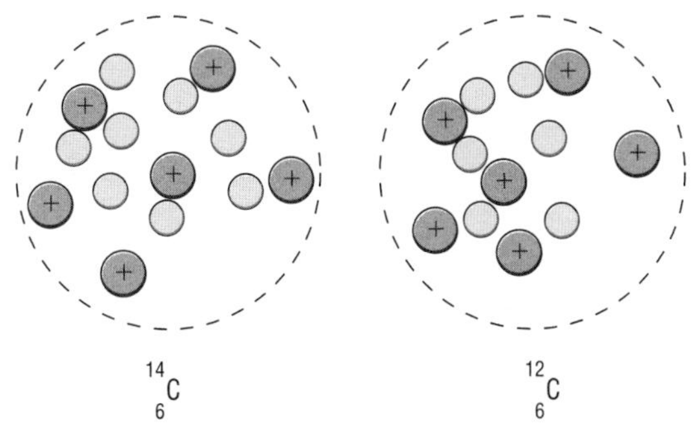

Figure 3 $^{14}_{6}C$ and $^{12}_{6}C$ nuclei

Worked example

Deduce the number of protons, neutrons and electrons in an atom of $^{14}_{6}C$.
The number of protons is given by the atomic number, so there are 6 protons.
In a neutral atom there are equal numbers of electrons and protons, so there are 6 electrons.
The total number of protons + neutrons is 14, and there are 6 protons. So there must be (14–6) or 8 neutrons.

Isotopes

Atoms which have the same number of protons but different numbers of neutrons. Examples include $^{14}_{6}C$ and $^{12}_{6}C$; $^{235}_{92}U$ and $^{238}_{92}U$.

In discussions of isotopes, the atomic number is often omitted – so $^{235}_{92}U$ can be referred to as ^{235}U or uranium-235.

Examiner tip

You will need to be able to deduce the numbers of each type of particle in a nucleus from a nuclear symbol.

Questions

1 **(a)** $^{235}_{92}U$ and $^{238}_{92}U$ are described as being isotopes of uranium. What is the meaning of the term 'isotopes'?
   (b) Calculate **(i)** the number of protons; **(ii)** the number of neutrons in each of these two isotopes.

2 **(a)** A proton is identical to the nucleus of which atom?
   (b) What was the evidence that Rutherford used to suggest that all atoms might contain protons?

3 Polonium-210 has the symbol $^{210}_{84}Po$.
   (a) (i) Calculate the number of protons and neutrons in a nucleus of a polonium atom.
   (ii) What are the actual masses (in kg) of a proton and a neutron?
   (iii) Use your answers to parts **(i)** and **(ii)** to calculate the mass of a polonium-210 atom.
   (b) (i) Polonium atoms are normally neutral. How many electrons will there be in a neutral polonium atom?
   (ii) Calculate the mass of the electrons in a polonium atom.

2.4 (5) Radioactivity

How science works

In this spread you will consider the risks of exposure to ionising radiation from radioactive sources (HSW 6a).

Radioactive decay

This eventually happens to all atoms containing an unstable nucleus – it 'decays' and emits radiation. The original (parent) nucleus is transformed into a new (daughter) nucleus.

In spread 2.4.3 you read that the discovery of radioactivity played a key part in the development of our model of the atom. Once the Rutherford model of the nuclear atom had been developed, it was possible to understand what happens when **radioactive** atoms undergo **decay**, giving out energy in the form of electromagnetic waves or particles.

Emissions from radioactive atoms

The three types of radiation emitted by unstable nuclei are known as alpha (α), beta (β) and gamma (γ) radiation. All three types of radiation possess enough energy to cause chemical changes to atoms and molecules – for example they can ionise (remove electrons from) atoms or molecules, as well as break chemical bonds. Because of this they are often described as *ionising radiations*.

Identity of the three types of ionising radiation

The three types of radiation have very different identities, as shown in Figure 1 and Table 1.

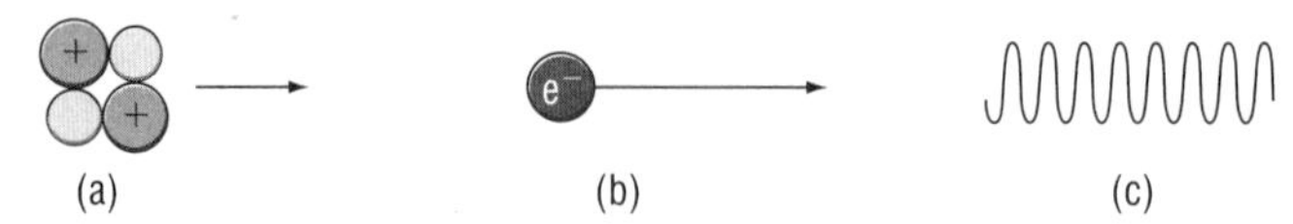

Figure 1 Ionising radiation: **a** α particle; **b** β particle; **c** γ radiation

It may seem surprising that an electron can be emitted from the nucleus of an atom, but scientists now know that in many unstable nuclei a neutron can decay into a proton and an electron – the proton stays in the nucleus and the electron is emitted as a β particle.

Properties of the three types of ionising radiation

Because the three types of radiation have different identities, the masses, charges and energy of each type of radiation are different. This means that some of the properties, which are measurable by experiments, are very different as well. Table 1 shows the properties of the three types of radiation.

Property	Type of emission		
	Alpha (α)	Beta (β)	Gamma (γ)
Relative charge	+2	−1	0
Relative mass	4	0.000 55	0
Range in air	Few centimetres	Few metres	Very long
Stopped by	Paper	Aluminium foil	Thick lead sheet
Ionising power	High	Moderate	Low

Table 1

Radiation dose

A measure of the effect which radiation has on cells. It varies depending on the energy of the radiation, the length of exposure and the type of cell exposed.

Biological hazards

As explained above, radiation is hazardous because of its ability to ionise molecules and to break chemical bonds. If living tissue is exposed to even a small **dose** of radiation then damage to the DNA molecule can cause mutations. These in turn may cause cancers or create genetic defects which are passed on to future generations. In larger doses, radiation can cause radiation sickness, radiation poisoning or result in burns to skin and other tissues.

Contamination and irradiation

Because alpha particles have the greatest ionising power they have the greatest potential to cause biological damage. Fortunately, alpha particles are stopped by air, clothing and thin plastic gloves. As a result, **irradiation** by alpha particles (exposure to alpha particles emitted from sources outside the body) does not present a great biological hazard. However if **contamination** occurs (if an alpha particle source enters the body) then because the cells are now in close proximity to the source significant damage can occur. This could happen, for example, if dust contaminated by radioactive atoms is breathed in, or if contaminated food and drink is consumed.

Beta particles and gamma radiation are not so easily stopped but they have less ionising power. Irradiation by beta particles is likely to be hazardous to tissues close to the surface of the body, such as the skin. Gamma irradiation could potentially damage cells anywhere in the body if the energy of the source is high enough. This is made use of in cancer treatment where the gamma radiation can be targeted at cancer cells deep within the body.

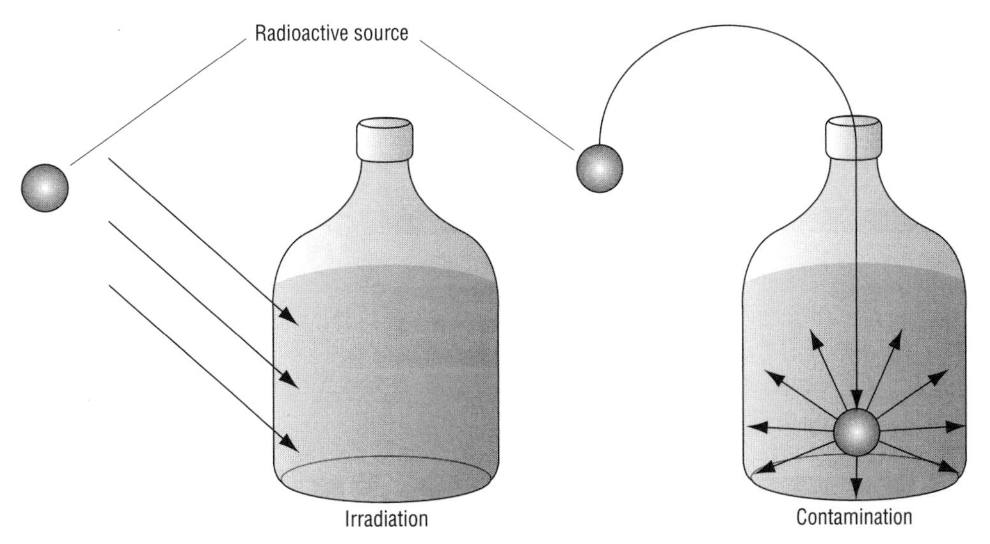

Figure 2 Comparison of irradiation and contamination

Activity

The annual dose of radiation received by individuals in the UK varies greatly depending on location, lifestyle and the number of medical examinations involving X-rays. Find out about the size of some of the doses produced by various activities, such as flying or eating certain foods. There are a variety of dose calculators available for estimating these doses (unfortunately many only cover the USA).

Irradiation

The exposure of an area or object to radiation. Although ionisation and other chemical processes may occur, the object does not become radioactive as it does not contain a source of radiation.

Contamination

The spreading of a radioactive source into a particular area or object. The object itself become radioactive as it now contains a source of radiation.

Questions

1 Eating seafood and exposure to medical X-rays are two examples of activities or lifestyle factors that may increase your total annual dose of radiation.
   (a) Eating seafood increases your dose by contamination; exposure to X-rays does so by irradiation. Explain the meaning of these words.
   (b) List two other activities or lifestyle factors which could increase your total dose.

2 Domestic smoke alarms contain small amounts of americium-241. This emits alpha particles and small amounts of gamma radiation.
   (a) Which of these types of radiation will penetrate through the plastic casing of the smoke alarm?
   (b) Why is low energy gamma radiation not considered to be a major risk factor?

2.4 (6) Nuclear processes

In spread 2.4.5 you looked at the radiation produced by radioactive decay. Here you will focus on what happens to the nucleus during decay and look at some other nuclear processes.

How science works

In this spread you will consider the possible sources of hazardous radiation (HSW 6a).

Radioactive decay

Because particles are lost from the nucleus during decay, the atomic number and mass number change after an atom has decayed.

Alpha decay

Two protons and two neutrons are lost during α decay – so the atomic number goes down by two and the mass number goes down by four.

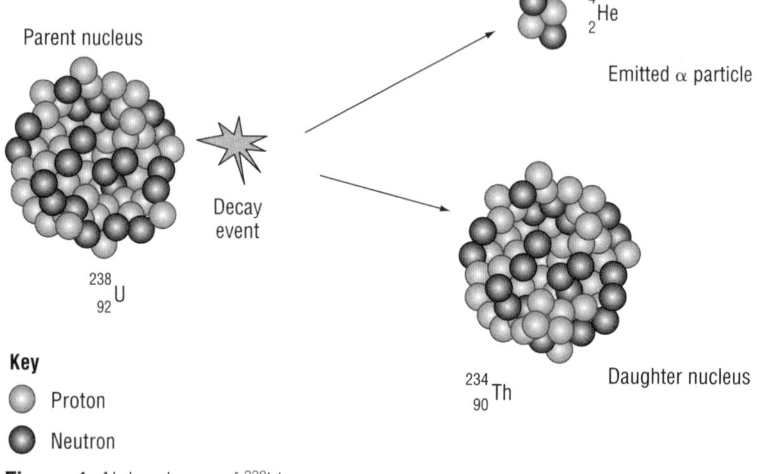

Figure 1 Alpha decay of ^{238}U

Beta decay

A neutron is transformed into a proton plus an electron – the electron is emitted. So the mass number does not change but the atomic number increases by one.

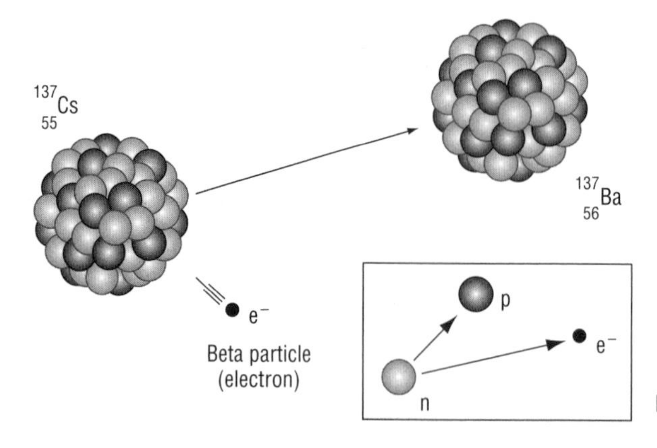

Figure 2 Beta decay of ^{137}Cs

Examiner tip

You may have to complete nuclear equations for decay processes. The key point will be to check whether the process is α decay or β decay. Then, so long as you know the symbol for the α or β particle produced, you will be able to follow some simple rules to balance the equation:

- α particle = $^{4}_{2}He$
- β particle = $^{0}_{-1}e$

Completing nuclear equations

1 Complete this nuclear equation for the alpha decay of a thorium atom:
$^{230}_{90}Th \rightarrow ^{?}_{?}Ra + ^{4}_{2}He$
- The mass numbers must balance (the top row of numbers) – so the missing mass number is 226 (226 + 4 = 230).
- The atomic numbers must balance (the bottom row of numbers) – so the missing atomic number is 88 (88 + 2 = 90).

2 Write a nuclear equation for the beta decay of the unstable nucleus $^{90}_{38}Sr$ into a nucleus of yttrium (Y).
- First you need to write down what you know about this process. A beta particle is emitted, so you can write $^{90}_{38}Sr \rightarrow ^{?}_{?}Y + ^{0}_{-1}e$.
- The mass numbers must balance (the top row of numbers) – so the missing mass number is 90.
- The atomic numbers must balance (the bottom row of numbers) – so the missing atomic number is 39. Be very careful with the –1 charge of the beta particle: 39 + (–1) = 38.

Nuclear fission

Radioactive decay is an entirely natural process and the rate at which it happens cannot be affected by any man-made effect. However **nuclear fission** is a process that is triggered artificially. It was first successfully carried out in 1939 and, just six years later, provided the energy source for the atomic bombs dropped on Japan at the end of the Second World War.

Because the products of fission include high-energy neutrons, a single fission event can trigger other events, causing a chain reaction.

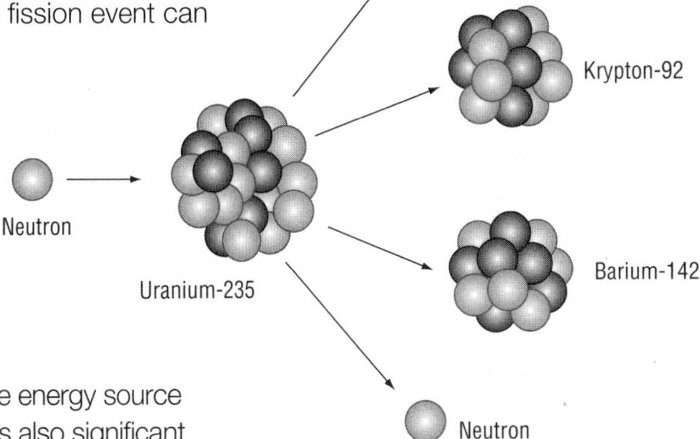

Figure 3 Fission of uranium-235

Nuclear fusion

Nuclear fusion is the process by which stars (including the Sun) produce massive amounts of energy. It is possible to cause this to happen artificially – it can occur uncontrollably in the thermonuclear reactions powering the hydrogen bomb. However, attempts to use fission in a controlled way to generate energy have not yet produced sustainable fusion reactions.

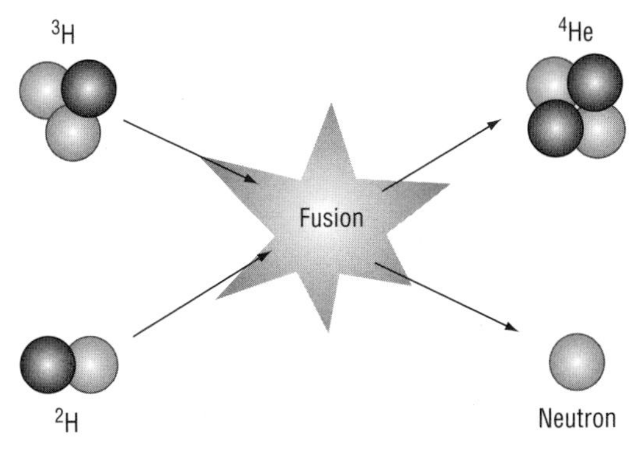

Figure 4 Fusion of ^{2}H and ^{3}H

As well as providing the energy source for stars, this reaction is also significant because it is the process by which heavier elements are produced. Most of the atoms in the Universe are hydrogen atoms – the elements which make up most of the Earth, such as carbon, oxygen and iron, were produced by fusion reactions which occurred in the extreme conditions of a supernova (an exploding star).

Nuclear fission

The splitting of a nucleus into lighter fragments, caused by the absorption of a neutron. Energy, in the form of gamma rays, other neutrons and heat are also produced.

Nuclear fusion

Two nuclei join together to produce a single heavier nucleus. Very high temperatures are required to cause this to happen, but large amounts of energy are released when fusion occurs.

Examiner tip

As with the equations for nuclear decay, you may be asked to complete equations for fission or fusion processes. Again, the key point is that the atomic numbers and mass numbers must balance on each side of the equation.

Activity

Find out about the world's only known *natural fission reactor*.

Sources of radiation

We are continually exposed to low levels of ionising radiation. This radiation comes mostly from natural sources and normally it is at far too low a level to be regarded as a health hazard.

Natural sources could include:

- gamma radiation from radioactive isotopes in underground rocks
- alpha and gamma radiation from radon – a radioactive gas (formed from the decay of uranium) that can sometimes seep to the surface
- cosmic 'rays' – high-energy particles such as protons and electrons emitted by stars (including the Sun), supernovas and maybe from distant black holes
- emissions from the very low levels of radioactive isotopes (e.g. potassium-40) present in food and drink.

Artificial sources could include any radioactive isotopes released when atomic weapons are tested, or from nuclear power station accidents. In addition, ionising radiation is used in a carefully controlled way in medical procedures such as X-ray investigations. You looked at how these are used in dose calculations in spread 2.4.5.

Questions

1 **(a)** Describe the differences between *decay*, *fission* and *fusion* – giving nuclear equations to illustrate your answer.
(b) One of these processes is described as an 'artificial' process. Which one?
(c) Explain why fission processes can cause chain reactions.
(d) Fusion processes produce large amounts of energy. Why else are they significant?

2 Complete nuclear equations for these decay processes:
(a) $^{221}_{87}\text{Fr} \rightarrow ^{?}_{?}\text{At} + ^{4}_{2}\text{He}$
(b) $^{131}_{53}\text{I} \rightarrow ^{?}_{?}\text{Xe} + ^{0}_{-1}\text{e}$

3 Complete nuclear equations for these fission and fusion reactions:
(a) $^{2}_{1}\text{H} + ^{?}_{?}\text{He} \rightarrow ^{7}_{3}\text{Li}$
(b) $^{235}_{92}\text{U} + ^{1}_{0}\text{n} \rightarrow ^{?}_{54}\text{Xe} + ^{90}_{?}\text{Sr} + ^{1}_{0}\text{n} + ^{1}_{0}\text{n}$

2.4 (7) Options for future energy generation (I)

How science works

In this spread you will consider the possible risks of using nuclear fission as a way of generating energy (HSW 6a).

In spread 2.4.2 you saw that around 80–85% of the world's energy demands are met by burning fossil fuels. It can sometimes be difficult to analyse data from different sources because the units used for energy are often different (many are given in 'barrels of oil equivalent'). However, it can be estimated that the total energy consumed worldwide is currently about 5×10^{20} J per year.

Remember, of course, that although we often talk about energy produced, generated or consumed, in reality it is only being transferred from another form – the chemical energy stored in fossil fuels, the electromagnetic energy of solar radiation or the energy stored in the nucleus of an unstable radioactive atom such as ^{235}U. When we use these sources to 'generate' energy, the energy is transferred into a useful form – normally heat, which is then used to drive turbines (to produce electrical energy) or to power vehicles.

To consider how the ways in which the world generates usable energy might change in future we have to look at a number of issues. First we will consider the so-called non-renewable energy sources – fossil fuels and nuclear fuels.

How long will our reserves last?

It is very difficult to estimate exactly how much fossil fuel remains in the Earth's crust. There are similar problems in estimating the amount of mineable nuclear fuel (such as uranium). One estimate puts the amount of energy stored in these sources as:

- fossil fuels (oil, coal, natural gas) 4×10^{23} J
- nuclear fuel 2.5×10^{25} J.

Activity

Calculate how long these reserves will last, assuming present rates of energy use. Do you think these estimates are realistic?

However, some scientists believe that it may be possible to extract vast amounts of methane (natural gas) from structures called methane hydrates which are found in sediments on the sea bed. If these reserves are included, the estimate of fossil fuel reserves increases to as much as 6×10^{24} J.

The increasing cost of energy

As the reserves of fossil fuels dwindle, and the technology required to extract the fuel from the ground becomes more expensive, it is inevitable that the price of these energy sources will increase. This may make their use uneconomic, and start a shift away from fossil fuel use towards other sources of energy.

Competition for crude oil

Another factor that makes the use of crude oil as an energy source less desirable in the future is that crude oil is also the most important raw material for the chemical industry. Most of the polymers (plastics) that we use in our everyday lives are derived from crude oil, as well as paints, detergents and pharmaceuticals. Furthermore, in most cases there is no alternative raw material – so crude oil may need to be reserved as a source of chemicals rather than of energy in the long term.

How does energy generation affect the environment?

In spread 2.4.2 you looked in detail at the environmental effects of burning fossil fuels. On top of this, there are also the environmental effects associated with mining, pumping or extracting the fuel from the Earth's crust – as well as of transporting the fuel to where it is needed.

Nuclear fuel, such as uranium, must also be mined and transported. However, the main environmental concern is, of course, the biological hazard of the radioactivity of the fuel and its products.

Nuclear power – the risks

Controlled fission reactions take place in nuclear power plants. The energy produced is transferred into heat energy, which is used to drive turbines. In order for fission to be possible, the natural mix of isotopes in uranium must be altered so that there is a greater proportion of the radioactive ^{235}U isotope. Both the fuel and the products of fission are therefore highly radioactive, and leaks or explosions will release these into the environment.

Figure 1 Dumped equipment next to the Chernobyl nuclear power plant

Applications: Chernobyl

In 1986 an explosion in a reactor at Chernobyl, in Ukraine, caused a cloud of radioactive material to be released. This drifted westwards and affected much of northern Europe. It is estimated that several thousand extra deaths from cancer might be attributable to exposure to radiation from Chernobyl.

Activity

Find out more about the causes and effects of the Chernobyl explosion.

Another major issue is the fact that the products of radioactive decay will remain radioactive for thousands of years because they have long **half-lives**. Because of the random nature of radioactive decay, it is never possible to say when the final atom will have decayed. However, a graph shows the way in which the activity of a sample decreases over time. Half-lives can range from a fraction of a second to billions of years. For example, the half-life of plutonium-239 is about 24 000 years. This isotope of plutonium is a product of the fission of uranium but it can also be recycled as a fuel in other designs of nuclear reactors.

Half-life

The time taken for the number of atoms of a sample of a radioactive isotope to fall to a half its original value.

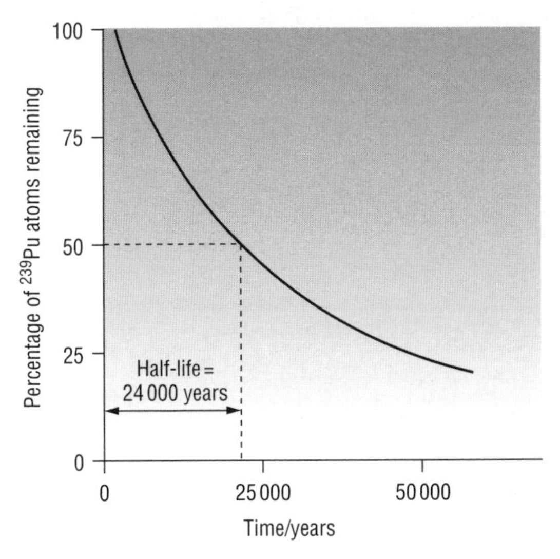

Figure 2 Decay graph of ^{239}Pu showing its half-life

To ensure that the products of fission, such as ^{239}Pu, do not pose a biological hazard it necessary to find a way of storing these waste products of nuclear fission safely. This means that:

- the emissions (mostly alpha and gamma) must be absorbed in the storage device
- there must be no danger of the ^{239}Pu leaking out to contaminate, for example, groundwater
- the container must be able to withstand thousands of years of exposure to heat, water and earth movements, such as earthquakes.

Question

1 Write a short essay discussing the advantages and disadvantages of generating energy from nuclear fission compared to the use of fossil fuels.

2.4 (8) Options for future energy generation (II)

How science works

In this spread you will evaluate some of the alternative strategies being proposed for energy generation (HSW 6a). You will also be introduced to the method of peer review which is used to validate new and possible radical new ideas, such as Cold Fusion (HSW 7a, 7b).

Activity

Calculate what percentage of the Sun's energy we would need to trap in order to meet the total yearly global energy consumption (see spread 2.4.7).

Is all the energy received by the Earth able to be trapped in this way? What fraction do you think could be trapped in this way?

As long-term energy generation options it is clear that the use of both fossil fuels and nuclear fission to meet our energy needs has major drawbacks. Unless we can develop new technological solutions to the problems outlined in spread 2.4.7 we may need to rely on alternative sources of energy much more.

Renewable energy

These make use of natural sources of energy, such as sunlight, wind, waves and tides. All these derive their energy from processes which occur away from the Earth:

- energy from the Sun produces heat which drives the winds and waves
- the movement of the Moon causes the tides.

As a result, these sources of energy are continually being replenished. It is calculated that the Earth receives about 4×10^{24} J of energy from the Sun every year.

Solar energy

Cheap, simple solar panels are now available to make use of the Sun's energy. They can be used to simply heat water for hot-water systems, or they can be used to generate electricity.

Wind energy

Large turbines, often arranged to make large 'wind farms', are increasingly being seen in suitably exposed land areas, or out at sea. The movement of the turbine generates an electric current.

Figure 1 Wind farm

Wave and tidal power

These both rely on the movement of water, whether caused by winds or by tides. This movement is then used to drive a turbine, generating an electric current.

Disadvantages

At present, the use of these renewable sources makes up only a tiny fraction of the total global energy generation. There are several disadvantages of these techniques:

- In order to generate sufficient energy, very large numbers of turbines or panels have to be used – many people claim this will create 'visual pollution'.
- Many parts of the world do not have suitable climates for solar or wind energy, nor suitable coastlines for wave and tidal energy.

- Wave and tidal power stations cause disruption to marine ecosystems – for example tidal barrages across estuaries affect the habitat of wading birds; wind turbines are said to pose a risk to birds which might fly into them.
- Solar, wave and wind power are unreliable and unpredictable, depending as they do on weather patterns.

At the moment, the price of electricity generated by these methods is significantly higher than that made using non-renewable energy resources. However, political pressure may cause this to change.

Nuclear fusion

If hydrogen nuclei could be made to undergo controlled nuclear fusion, as described in spread 2.4.6, then the world could make use of an almost unlimited fuel source – hydrogen can be obtained easily from water – and a nuclear reaction in which there is very little radioactive waste. For 40 years, many scientists have dreamed of being the first to generate energy from a fusion reactor successfully. However, the problems are enormous – in order to fuse, the nuclei must be heated to a temperature of up to 100 000 000 °C. The incredibly hot gas is known as a *plasma* and is contained in an incredibly strong magnetic field (magnetic fields are covered in spread 2.4.11 *Electrical and magnetic fields*).

Although fusion has been observed to happen, it can only be maintained for a few seconds. The energy required to generate the conditions for fusion is much greater than the energy that has been obtained from fusion. The next ten years' research may well be decisive in determining whether or not fusion is a viable future technology.

Figure 2 A fusion torus

Questions

1 (a) List three forms of renewable energy which make use of (directly or indirectly) energy received in the form of electromagnetic radiation from the Sun.
(b) Name a form of renewable energy which does not rely on energy from the Sun.

2 Nuclear fusion has been claimed to be a safe and sustainable form of energy, compared to nuclear fission.
(a) Describe two advantages of nuclear fusion processes compared to nuclear fission.
(b) Explain why, despite these advantages, there is currently no prospect of a workable nuclear fusion reactor.

Applications: biofuels

Some of the fuels used in transport, such as diesel and petrol, could be replaced by substances such as ethanol and 'biodiesel'. These can be produced from crops such as sugar cane or soya, so they are 'renewable'. Although carbon dioxide is still produced when the fuel burns, the crop absorbs the same amount of carbon dioxide from the air when it grows. So it is claimed that the use of biofuels is 'carbon neutral' and will not cause climate change.

However, in order to replace fossil fuels for transport energy use, large areas of land currently used for growing food would need to be converted to growing biofuel crops.

Cold fusion

If you had read the newspaper headlines in March 1989, you might have believed that the world's energy generation problems could just be over. Two scientists – Stanley Pons and Martin Fleischmann – called a press conference to announce that they had successfully caused nuclear fusion to happen using a simple laboratory electrolysis apparatus. Using heavy water – formed using deuterium nuclei (^2_1H) – and a palladium electrode, they claimed to have produced enormous amounts of heat.

But Fleischmann and Pons has made their announcement without exposing their results to **peer review** – where other scientists who are experts in the same field evaluate the work and decide whether the conclusions are valid. Only after this process has happened are the results of scientific research published.

In fact when other scientists tried to replicate the work of Fleischmann and Pons, they were unable to produce the same results and the theory of cold fusion, as announced in 1989, is now discredited.

But puzzling anomalies still remain in some similar experiments and work continues to investigate whether fusion is possible without a huge input of heat energy.

2.4 (9) Electricity transmission and distribution

Power

The rate at which energy is transferred – the unit of power is the watt (W). A power of 1 W means that 1 J of energy is transferred every second:

$1\,W = 1\,J\,s^{-1}$

In previous spreads you have seen that electricity is generated from a number of energy sources – for example the burning of fossil fuels or the process of nuclear fission. The heat released in these processes is used to generate steam, which is then forced through turbines causing them to turn. It is the energy of this turning process which is eventually transferred into electrical energy. In this spread it will be more helpful to refer to the generation and distribution of power rather than simply 'energy'.

The UK, like all modern societies, relies on a network of power lines to transmit and distribute this power to consumers.

Electricity transmission

The delivery of electrical power from power station to a sub-station.

Electricity transmission and distribution

Figure 1 shows part of the system used to transmit and deliver power from power stations to consumers.

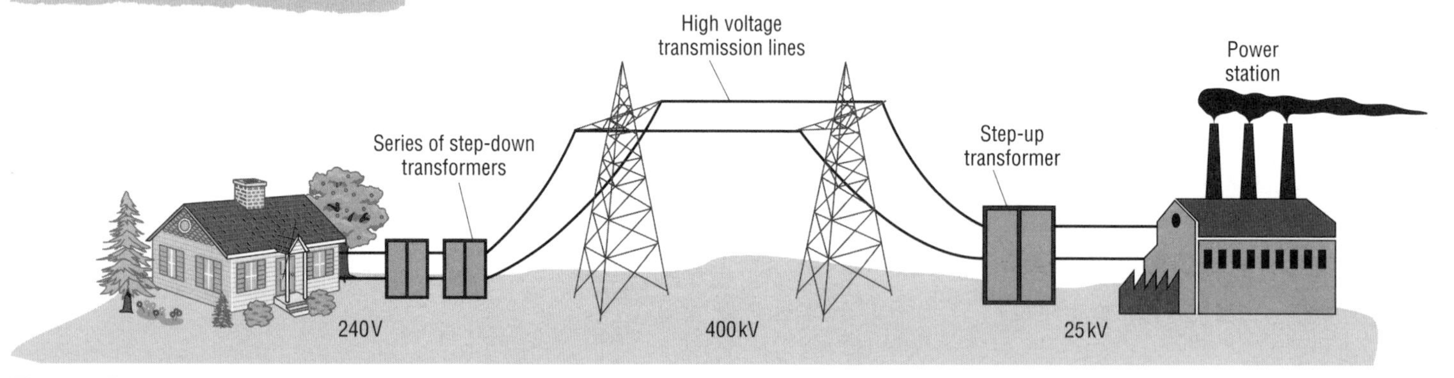

Figure 1 Power transmission and distribution

Electricity distribution

The delivery of electrical power from a sub-station to consumers.

In the UK, the transmission lines form a network, known as the national grid, that allows power to be distributed between one point and another by a number of alternative routes.

To understand the processes involved in the transmission and distribution system, you will find it helpful to be clear about some of the key terms.

Alternating current (a.c.)

Power stations generate an alternating **electrical current** – this means that the direction of flow of the electrons carrying the charge alternates many times every second. Alternating current is easier to generate from turbines and also allows the use of transformers to change the **voltage** of the electricity.

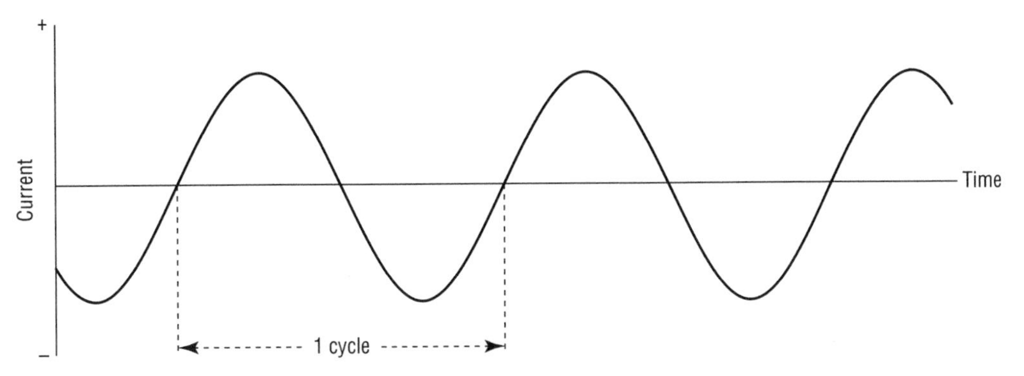

Figure 2 Graph to show how current varies in a 50 Hz alternating current supply – there are 50 complete cycles every second, which means the current changes direction 100 times a second

Transformers

These are devices which can change the voltage of alternating supply. Step-up transformers increase the voltage; step-down transformers decrease the voltage.

In the national grid transmission system, a step-up transformer is used to increase the voltage to a very high value (e.g. 400 kV) in the high-voltage power lines. A series of step-down transformers are used in the distribution system to bring the voltage down to 240 V, which is the voltage used in our domestic supply.

Electrical current

A flow of charge – in the case of metals, the charge will be carried by electrons. The unit of current is the amp (A) where 1 A means that 1 C of charge flows in one second:

$1\,A = 1\,C s^{-1}$

Current can be calculated using:
current = charge / time

Voltage

Also known as 'potential difference', this is a measure of the amount of energy that can be transferred when a charge moves between two points. The unit of potential difference is the volt (V) where 1 V means that 1 J of energy is transferred by each coulomb of charge:

$1\,V = 1\,J C^{-1}$

Questions

1 **(a)** The current in electrical power lines is described as 'a.c.'. Explain what this means.
  (b) The frequency of this a.c. supply is 50 Hz. State the meaning of the term 'frequency' (see spread 1.1.1).

2 **(a)** If the current flowing in a simple electrical circuit is 0.4 A, calculate the charge that will flow around the circuit in 2 minutes (120 s).
  (b) This amount of charge transfers 480 J when it passes around the circuit.
    (i) How many joules are transferred in 1 second?
    (ii) Hence, write down a figure for the power of this electrical circuit. Give the appropriate unit.

3 **(a)** Suggest why a grid system is used to distribute electrical energy from power stations to towns and cities.
  (b) Describe two ways in which transformers are used in electrical transmission and distribution systems.

2.4 (10) Electrical circuits

Direct current (d.c.)

Current in which the flow of charge is in one direction.

It is important to realise that, even on the enormous scale of the national electricity supply network, the system must be thought of as a series of complete circuits through which charge can pass. The network uses a.c. but we will introduce the key ideas about electrical circuits by thinking about direct current (d.c.) which makes explanations slightly easier. However, the equations you meet can still be applied to the much larger scale of electrical transmission networks.

Using ideas about current and voltage

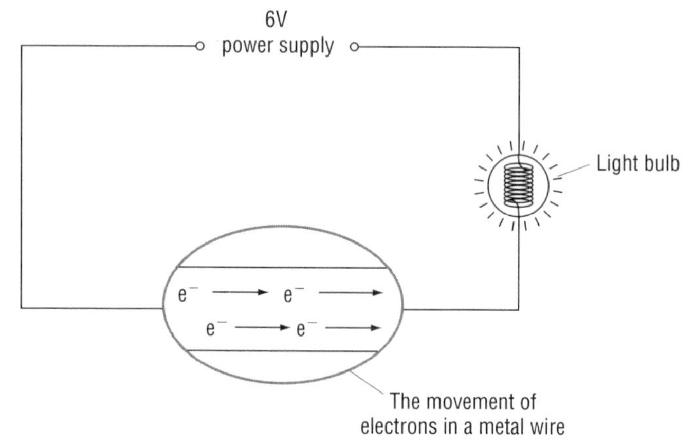

Figure 1 Simple electrical circuit showing a 6V power supply – the magnified section shows moving electrons in the wire

Examiner tip

Diagrams showing the direction of flow of charge can sometimes be confusing because they often show the direction in which positive charge will flow, even though the electrons that actually carry the flow are negative. You will not need to worry about this distinction in this course.

In circuits like that shown in Figure 1, the power supply transfers energy at a certain rate to the electrons. These move around the circuit as an electric current transferring their energy into other forms – in this case heat and light in the bulb. The filament in the bulb is said to have a high **resistance**.

The power of a circuit (the rate at which energy is transferred) can be calculated using power = voltage × current

Resistance

The ability of a material to resist the flow of electrical charge – the unit of resistance is the ohm (Ω).

Energy is transferred when current flows through a resistance. The resistance can be calculated using Ohm's law:

resistance = voltage / current

Worked example

A 240V power supply is used in a circuit with a 100W bulb.

(a) Calculate the current in the bulb.
Power = voltage × current, so current = power / voltage = 100 / 240 = 0.42A

(b) Use this figure to calculate the resistance of the bulb.
Resistance = voltage / current = 240 / 0.42 = 571 Ω

Understanding the design of the transmission system

As you saw in spread 2.4.9, one of the most important aspects of the national electricity transmission system is that the voltage is stepped up to a very high value in the power lines. This is done to reduce loss of power over the long distance that the electricity must travel. Power loss depends on two things:

- the size of the current
- the resistance of the wire.

In fact, power loss = $(\text{current})^2$ × resistance. Even though power lines are designed to have as low a resistance as possible, a high value for the current will mean huge losses of power. So the voltage is stepped up. Because power = voltage × current, a higher voltage means that the same power can be supplied at a lower current – this reduces the power loss.

Examiner tip

In calculations on the electricity supply system, the numbers are likely to be very large, with a lot of use of the prefixes you met in spread 1.1.4. Make sure that you are familiar with these – and with using your calculator to handle very large numbers.

Worked example

A power station provides 500 MW of power to a transmission line at a voltage of 400 kV. The total resistance of a length of transmission line is 40 Ω.

(a) Calculate the current in the transmission line (assuming that the same power is supplied).

Current = power / voltage = $500 \times 10^6 / 400 \times 10^3 = 1250$ A

(b) Calculate the power loss in this length of transmission line.

Power loss = (current)^2 × resistance = $1250^2 \times 40 = 6.25 \times 10^7$ W

This represents 12.5% of the total power transmitted. If the voltage was not stepped up, the current would be greater and the power loss also much greater – because of the factor of (current)^2 in the equation.
The voltage must, however, be stepped down when it is distributed to consumers for safety reasons.

The safety of high-voltage supplies

Close approach to high-voltage power lines carries the risk of electrocution – where a high potential difference across the body causes a massive current to pass through it. Power cables are suspended at a height of 8 m or more to ensure that it is impossible, under normal circumstances, to get close enough to the cable for this to be a possibility. In some areas of the UK – for example in areas of great natural beauty – power cables are routed underground. Because soil and rocks can act as conductors, these cables must be surrounded by a plastic insulator.

However, concerns still exist about the possible adverse health effects of the electromagnetic fields which surround these cables – this is discussed in spread 2.4.11.

Figure 2 An electricity pylon

Questions

1 **(a)** Explain why high voltages are used in national grid electrical transmission lines.
(b) Describe one important safety feature of such lines.

2 In a simple d.c. circuit, a 12 V power supply is used. The current flowing around the circuit is 0.02 A. Calculate the total resistance in the circuit. Give the appropriate unit.

3 The power transmitted in a 400 kV transmission cable is 160 MW. Calculate the current in the cable.

4 **(a)** Calculate the power transmitted in a 220 kV transmission cable if the current is 800 A.
(b) **(i)** If the resistance in a 10 km length of this power line is 2.75 Ω, calculate the power loss in this length of cable.
(ii) What percentage of the total power transmitted is lost in this 10 km length?

2.4 (11) Electrical and magnetic fields

Even though high-voltage power cables are suspended well above the ground, it is still possible for the current passing through the cable to produce an effect at ground level. This is because electric currents produce an electromagnetic field. Some researchers have claimed that these fields could have an adverse health effect on people or animals regularly exposed to them.

Electric field

A region of space around a charged object in which any other charged object may experience a force.

Magnetic field

A region of space around a magnetic pole in which any other magnetic pole may experience a force.

Field line

A line which shows the direction of movement of a + charge in an electric field, or a north pole in a magnetic field.

Fields

You will be familiar with electrostatic forces – for example the attraction of a positive and negative charge for one another. Similarly you will also have seen magnets attracting or repelling each other – for example the north pole of one magnet will attract the south pole of another. These effects can be explained by the idea that charges and magnetic poles create **fields** around themselves. Other charges, or poles, which enter these fields also have fields around them and forces are produced when these fields interact.

Field lines

Field lines provide a way of visualising these fields, and also predicting the direction in which forces will act.

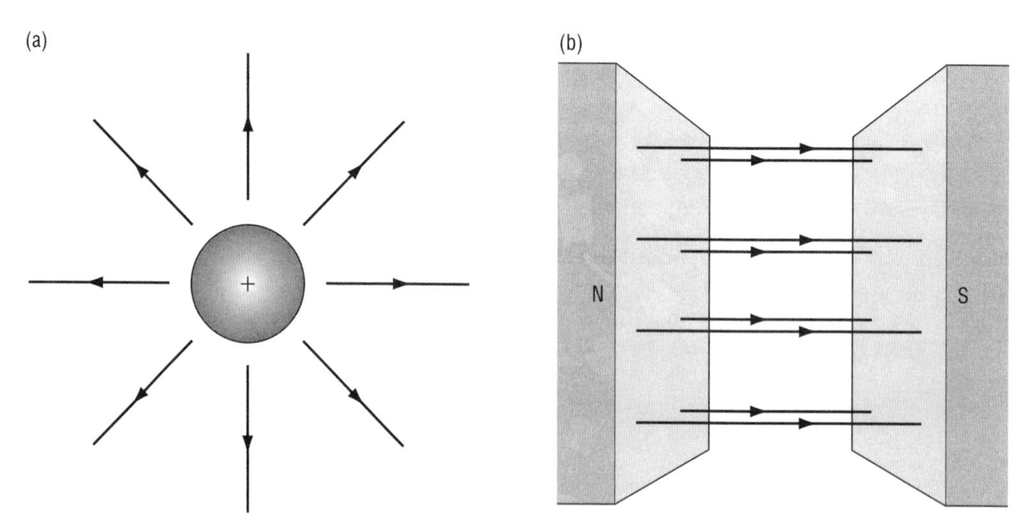

Figure 1 Field line patterns: **a** around a positive charge; **b** between the poles of magnets

Examiner tip

You do not need to remember these field line patterns (or the more complex patterns caused by electrical currents) but you may need to interpret them in terms of the direction and size of the forces they produce.

Field lines show the direction of the force produced on other objects in the field, and also indicate the strength of the field – and hence the strength of the force. When field lines are close together the field has a high *flux density* – the strength of the force is proportional to the flux density.

Fields and electric currents

Alternating magnetic fields

Moving charges (electric currents) – such as in a high-voltage power cable – also cause magnetic fields. A current flowing in a wire causes a magnetic field around the wire – the pattern is shown in Figure 2.

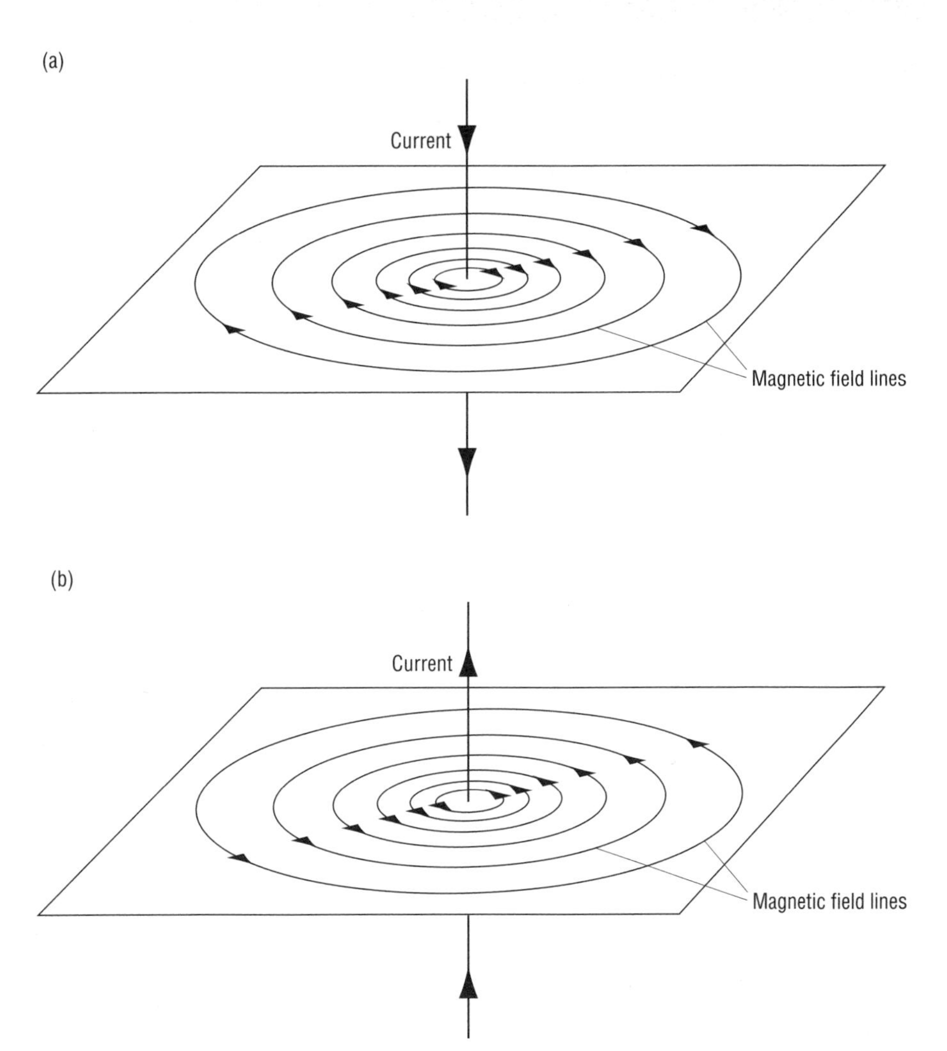

Figure 2 Magnetic field: **a** around a current-carrying wire; **b** with the direction of current reversed

Notice how the direction of the field is reversed if the current is reversed. This is what happens in a high-voltage transmission cable carrying an alternating current – the direction of current reverses 100 times a second. So the direction of the field reverses 100 times a second (50 Hz). Any charged objects exposed to this alternating field will experience a force alternating in direction at the same frequency, causing them to move – so small alternating currents would result.

Induced electrical fields

In the late nineteenth century, it was discovered that there was a link between electric and magnetic fields – they could be thought of as a single linked field, known as an electromagnetic field. One of the implications of this is that if there is an alternating magnetic field in a region of space then an alternating electric field will be *induced* in the same region. So there are also alternating electric fields around power lines.

Questions

1 Field lines are used to represent the magnetic field around a current-carrying wire (for example in Figure 2).

(a) What information does the closeness of the lines provide?

(b) What does the arrow on the lines tell you?

(c) How many times in one second does the magnetic field change direction?

2 A current flowing in a wire will cause a magnetic field around a wire. Why are there sometimes electrical fields around current-carrying wires?

2.4 (12) Alternating fields and epidemiology

How science works

In this spread you will consider the possible health risks of the alternating fields produced by power lines and consider the difficulty of assessing the risks (HSW 6a, 7a).

The presence of alternating fields around high-voltage power cables created concerns in the 1970s when some scientists suggested a link between exposure to these fields and diseases such as leukaemia and other cancers.

How strong are the fields?

Because the strength of the field reduces rapidly with distance, it is only people directly below the power lines who are likely to be exposed to significant field strength. However, several features of the power transmission system will reduce the strength of the fields still further:

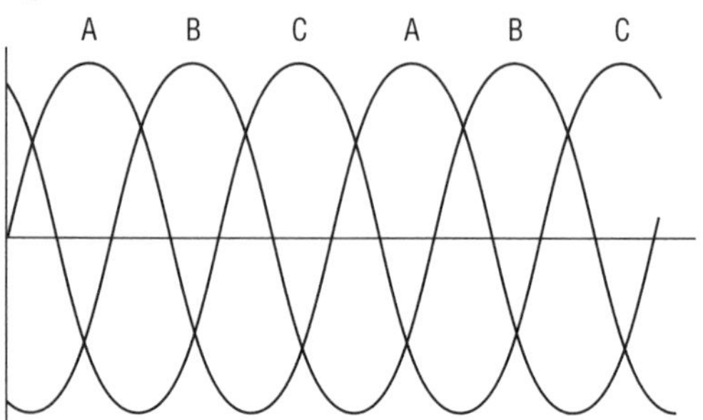

Figure 1 Cancellation of three out-of-phase alternating currents

- The current is kept as low as possible (by stepping up the voltage) – this reduces the strength of the magnetic field.
- The electrical power is transmitted by three sets of conductors in a three-phase supply. This means that the fields produced by the currents tend to cancel out – field lines produced by one conductor are in opposite directions to those of the other two.
- The power cables are kept high above the ground.

A typical field beneath a high-voltage cable might be 10 μT (the tesla (T) is the unit of magnetic field strength). For comparison, the strength of the Earth's magnetic field is about 40 μT.

The alternating electric fields produced beneath power lines might have a strength of about 1000 $V m^{-1}$. For comparison, the 'natural' electric field, due to charged particles in the atmosphere, is normally about 100 $V m^{-1}$. However, in a thunderstorm this may rise to 20 000 $V m^{-1}$ or more.

Possible biological effects

So the magnetic fields are lower than the 'natural' fields. However, the fields produced by high-voltage power cables are alternating, whereas 'natural' fields are unidirectional.

However, if the claims of possible health risks are to be believed, there would have to be a mechanism by which the alternating fields could cause cell damage. It has been suggested that heat generated by currents induced in cells may cause damage, or that ions produced in the air around the cables may have some toxic effect.

It is possible to test this by exposing cell cultures, or even animals, to alternating magnetic fields of different frequencies and strengths.

Applications: health hazards

The alternating electric field beneath power cables is normally thought to be so low that it cannot possibly be a hazard to human health. However, although the field is normally low, a combination of 'sagging' cables and bumpy ground can locally increase the field to much higher levels.

Figure 2 High local field strengths

Many farmers report that grazing cows tend to avoid these 'hot-spots'. Some mysterious deaths of cows have even been attributed to electrocution in these regions, but there is no reliable evidence to link these deaths with power cables rather than lightning.

Epidemiology

Because it is obviously not ethical to deliberately expose human beings to potentially damaging factors, scientists use **epidemiological studies** to investigate the link between alternating electromagnetic fields and human health. We will look at two common types of epidemiological studies. Both attempt to calculate the relative risk of developing a particular health risk after being exposed to a specific factor. In the case of exposure to high-voltage power cables, no clear evidence has yet emerged showing a link with human health.

Case–control studies

If, for example, it is thought that alternating magnetic fields could cause leukaemia (a cancer of the blood) to develop, then a case–control study could be carried out:

- A group of people suffering from leukaemia are selected.
- The past history of this group is then investigated to discover the level of exposure to alternating magnetic fields – for example if they have ever lived directly under a power line.
- A second group of people who are not suffering from leukaemia are selected. They are chosen so that they match the first group as closely as possible – by age, social class etc.
- The history of this second group is also investigated.
- If the first group are found to have been exposed to a greater level of alternating field than the second then it may indicate that there is a link between this exposure and the development of leukaemia.

The interpretation of the results can be very difficult and is often complicated by the fact that many factors that need to be controlled are related in a complex way – for example it may be that the standard of housing situated under power lines is poorer than housing situated much further away. There is also a much greater chance of bias – intentional or unintentional – in the selection of the two groups.

Cohort studies

Case–control studies work backwards to try to establish the link between a risk factor and a health problem. Cohort studies take the opposite approach and follow groups of people over many years to see if health problems emerge:

- A group of people are identified who have been exposed to a factor – for example they all lived under high-voltage power lines as children.
- A second group of people are identified, as similar as possible except that they have not lived under power lines.
- The two groups are then monitored for the development of, for example, leukaemia. This may continue for years, or indeed a whole lifetime.

The results are much less likely to be subject to bias, and can also provide a figure for the absolute risk of a health problem arising from exposure to a specific factor. However, it takes a very long time to produce results.

Questions

1 State three features of high-voltage transmission systems which reduce the strength of the magnetic fields beneath them.

2 Outline how you would carry out a study to try to investigate the relationship between smoking and lung cancer. Is the study you have described a case–control study or a cohort study?

3 Suggest why exposure to the very high electric fields present during a thunderstorm may not be harmful to human health, whereas much smaller fields due to power lines has been a cause of concern to some people.

Epidemiology

The science of epidemiological studies in which the causes of disease, or the way in which it spreads, are investigated.

Applications: other epidemiological studies

Epidemiological studies have been carried out to estimate the risks of exposure to ionising radiation. Many of these are cohort studies of inhabitants of Hiroshima and Nagasaki, who were exposed to known doses of radiation when atomic weapons were used against Japan in the Second World War. These data have been used to estimate the risks of much lower doses.

In recent years, concern about the use of mobile phones and exposure to radiation from mobile phone transmitter masts has also generated a range of epidemiological studies.

2.4 (13) Analysing epidemiological data

How science works

In this spread you will analyse and interpret data which has been used to try and establish a link between alternating fields and damage to human health (HSW 5b, 5c, 7a).

Figure 1 High-voltage power line close to housing

A number of epidemiological studies have been carried out to investigate the hypothesis that children who develop leukaemia are more likely to have been exposed to low frequency magnetic fields from power lines than other children. The results of many of the studies are inconclusive. However, one study has caused a great deal of interest because it seems to provide support for this hypothesis – it was carried out in Sweden by M. Feychting and A. Ahlborn in 1993.

Feychting and Ahlborn conducted a case–control study. Their sample consisted of all children under 16 years of age who had lived within 300 m of any 220 kV or 400 kV overhead power lines in Sweden between 1960 and 1985. There was a total of 127 383 children. The children were studied from the time they began to live in the area of a power line up until 1985.

There were 39 cases of leukaemia that were indentified and studied. A total of 554 controls were selected who had not developed any form of cancer. These controls were selected from the 127 383 individuals followed in the study – so they had also lived close to power lines. For each cancer case, the aim was to randomly select several controls of the same age and who lived in the same area, close to the same power line.

The study then attempted to calculate the magnetic field strength that the children were exposed to. Because the exposure was in the past, the researchers looked at records for the power (strictly speaking the 'load') transmitted by each power line and worked out the likely magnetic field in the house where each child in the two groups lived. They checked the method of calculation by directly measuring the field close to power lines, and proving that it closely matched the field predicted from knowing the power of the power line.

The data for the two groups could then be compared in a variety of ways. In one study the numbers of cases and controls exposed to high and low fields were compared; in another the numbers of cases and controls living very close to or much further away from a power line were compared.

The results of these two analyses are shown in Tables 1 and 2 – the results are given for four categories of children.

Group	Cases	Controls
1 Exposed to fields >0.2 μT	7	46
2 Exposed to fields <0.1 μT	27	475

Table 1

Group	Cases	Controls
3 Living within 50 m of a power line	6	34
4 Living beyond 100 m from a power line	26	431

Table 2

Questions

1 If the exposure to different levels of magnetic fields had no effect on the development of leukaemia, what would the theoretical ratio of controls : cases be in each category? (Hint: look at the numbers of cases and controls used in the whole study.)

2 Calculate the actual ratio of controls : cases in each category.

3 According to your figures, in which categories are the ratios of controls : cases similar to the theoretical ratio expected?

4 For the other two categories, calculate a figure which indicates how many times greater the chances are of developing leukaemia. (This is often known as the 'odds' ratio.)

5 Do you think that these results could be the result of chance, or are they likely to be statistically significant?

6 The study took great care to ensure that the effects of other factors were controlled as well. Suggest a factor which might also be associated with living close to power lines which must be controlled in this study.

7 The measurement of the magnetic fields by the team, checking if the fields matched the predictions they made from knowing the power in the cables, was done using electronic meters which did not display the measurements directly. The data were retrieved by downloading them into a central computer. Suggest why this precaution was taken.

8 Why was it necessary to obtain data about as many as 127 000 children in this study? Does this help to explain why there have not been enough studies done to allow scientists to give a conclusive answer to the question of whether exposure to low frequency magnetic fields increases the risk of leukaemia?

2.4 Module 4 – Summary

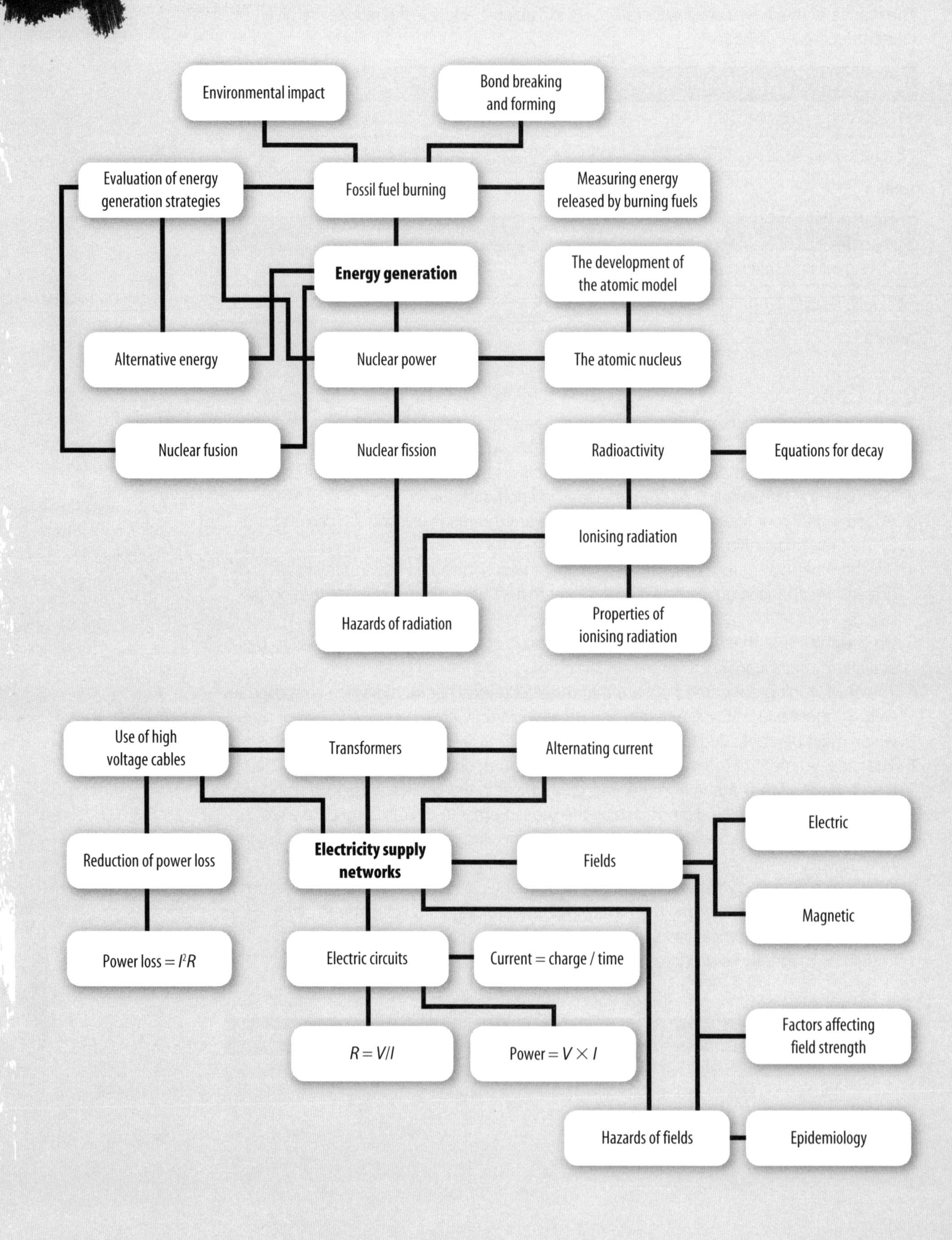

Practice questions

Low demand questions

These are the sort of questions that test your knowledge and understanding at E and E/U level.

1 **(a)** The burning of fossil fuels, such as coal, has a number of significant environmental effects. State three different environmental effects related to the burning of these fuels.

(b) Some data for the use of different sources of energy used in the USA are shown in the following table.

Energy source	Percentage
Coal	23%
Oil	40%
Natural gas	
Other	14%

(i) Calculate the percentage of energy needs supplied by natural gas.

(ii) Present this data in the form of a pie chart.

(iii) Suggest two forms of energy generation which could be included in the 'other' category.

2 Electrical power is generated in power stations and is then transmitted and distributed to consumers by a network of high-voltage power cables. The voltage used in these cables could be as high as 400 kV.

(a) Write 400 kV in volts.

(b) Why is it necessary to distribute electricity at such high voltages?

(c) Name the piece of equipment used to 'step up' the voltage from the power station to 400 kV in the transmission cable.

Medium demand questions

These are the sort of questions that test your knowledge and understanding at C/D level.

3 One way of comparing the amount of energy released by fuels is to use them to heat up water. In one experiment, two fuels – ethanol and hexane – were burnt. The heat released was used to heat 200 g of water in an aluminium can. The energy transferred to a mass of water is given by the equation:

energy (in J) = 4.2 × temperature rise × mass of water

0.50 g of ethanol produced a temperature rise of 19 °C

0.25 g of hexane produced a temperature rise of 17 °C.

(a) Calculate the amount of energy (in J) released by **(i)** ethanol; **(ii)** hexane in the experiment.

(b) Calculate a figure for the amount of energy released *per gram* for each of the two fuels.

(c) According to your figures, which substance would make the better fuel?

4 Carbon exists as several isotopes. One isotope, ^{14}C, is radioactive and emits β-particles when it decays.

(a) Explain the meaning of the term 'isotopes'.

(b) Copy the equation below, which shows the β-decay of a ^{14}C atom, and complete it by writing the appropriate missing numbers:

$$^{14}_{6}C \rightarrow {}^{?}_{?}N + {}^{0}_{-1}e$$

(c) Explain why the emission of β-particles by isotopes such as ^{14}C can be hazardous to human health.

High demand questions

This is the sort of question that tests your knowledge and understanding at A/B level.

5 One of the dangers of high-voltage power lines is that they give rise to an alternating magnetic field. Some scientists feel that exposure to such fields could be damaging to human health.

(a) **(i)** What is meant by the term 'magnetic field'?

(ii) Explain why the magnetic field around high-voltage power lines will be 'alternating'.

(b) The field produced by high-voltage power cables can be very high because the currents in such cables are very high. The equation which links power (W), current (I) and voltage (V) in an electrical circuit is $W = V \times I$.

(i) Use this equation to calculate the current in a 220 kV power cable which is used to deliver 360 MW of power.

(ii) If this power line delivered 360 MW of power for 6 hours, calculate the total energy which was supplied.

(iii) Describe two ways in which the design of power lines ensures that the magnetic field below the line is as small as possible.

Unit 1 Examination questions

1 The tropical rainforest ecosystem has a high net productivity. The mass of dry organic matter (biomass) produced per square metre is a measure of net productivity. Values for the annual net productivity of different trophic levels in a tropical rainforest ecosystem are given in Table 1.

Trophic level	Type of organism	Annual net productivity/g m^{-2}
1	Producers	2200
2	Primary consumers	310
3	Secondary consumers	44

Table 1

The producers in a tropical rainforest are mainly trees, which are often very large. One tree can be home to tens of thousands of insects which graze on the tree and to hundreds of larger organisms (e.g. frogs, birds and snakes) which prey on the insects.

(a) What is meant by the term 'ecosystem'? [2]

(b) **(i)** Write down one food chain that occurs in a tropical rainforest. [1]

(ii) Explain why a food web is more appropriate than a food chain as a description of the feeding relationships in an ecosystem. [1]

(c) Calculate the percentage of biomass in producers which is converted into biomass in primary consumers. [2]

(d) The annual net productivity of producers in temperate forests in the UK is about 1200 g m^{-2}. Give two reasons why the annual net productivity of producers is much greater in a tropical rainforest than in a temperate forest. [2]

[Total: 8]

(OCR AS Science 2841 Jun01)

2 Figure 1 shows the paths of two beams of electromagnetic radiation received by a remote sensing satellite. Beam A is visible light; beam B is thermal infrared radiation.

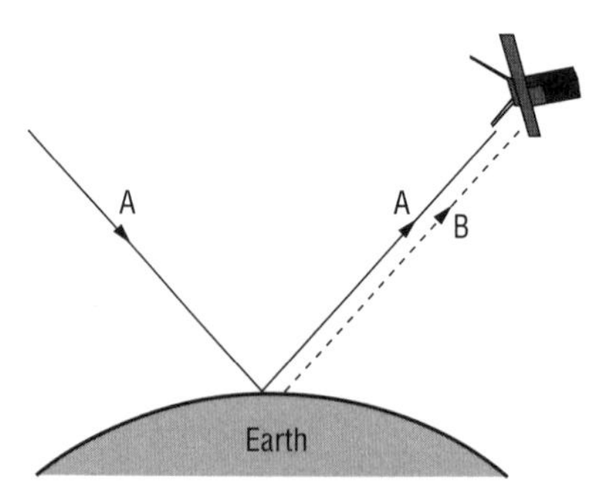

Figure 1

(a) **(i)** In what way does the wavelength of visible light differ from the wavelength of thermal infrared radiation? [1]

(ii) State one way in which the electromagnetic radiation in a wave of visible light and in a wave of thermal infrared radiation are the same. [1]

(iii) Name the region of the spectrum that lies between the visible region and the thermal infrared region. [1]

(b) Beam A is reflected by the Earth's surface. What two processes are involved at the Earth's surface to produce beam B? [2]

(c) Table 2 shows how land surface and clouds appear in daytime in remotely sensed images recorded in the visible and in the thermal infrared regions.

Type of radiation	Appearance of land	Appearance of clouds
Visible	Dark grey	
Thermal infrared	Pale grey	Black

Table 2

(i) Explain why the appearance of land is different when images are recorded using visible and thermal infrared radiation. [1]

(ii) Explain why clouds appear black in a thermal infrared image. [1]

(iii) What would the appearance of clouds be in the visible image? [1]

[Total: 8]

(OCR AS Science 2841 Jan02)

3 One of the problems of raising livestock is getting rid of animal waste. A common practice is to use water to wash the raw waste out of barns into lagoons. A lagoon is a big pit in the earth, sometimes lined with clay or a plastic liner. When a lagoon is full, the mixture is pumped out and sprayed onto fields as fertiliser.

(a) Name two plant nutrients in animal waste that makes it a good fertiliser. [2]

(b) Lagoons can leak. Animal waste can then find its way into surface waterways, killing fish and causing environmental damage. Leakage from the lagoon can cause eutrophication.

(i) What initial effect does eutrophication have on the growth of plant life in the waterways? Explain your answer. [2]

(ii) Explain how eutrophication results in the death of fish. [3]

(c) Application of synthetic fertilisers onto crops can also cause eutrophication.

(i) State one way in which synthetic fertilisers can enter surface waterways. [1]

(ii) State how pollution of surface waterways by these synthetic fertilisers can be reduced. [1]

[Total: 9]

(OCR A-level Science 2844 Jan05) omitting part c

4 Electromagnetic radiation can be used to transfer information. For example, visible light is used to produce images in both the human eye and a digital camera. A number of processes are involved in the formation of these images.

(a) Complete the phrases (i) to (iv) using the appropriate process from the following list:

Absorb diffract reflect refract

(i) Rods and cones in the human eye ____________ the light which falls upon them.

(ii) The edges of a photographic image can be blurred because light can __________ due to the small aperture used in the camera.

(iii) The lens in both a camera and the human eye causes the light rays to __________ and produces a sharp image.

(iv) Light rays from a source of light fall upon an picture which cause them to __________ into the human eye. [4]

(b) Human beings see in colour during the daytime, but in black and white when light levels are low, such as in moonlight. Explain this difference by discussing the different sensitivities of rod and cone cells. [3]

(c) Electromagnetic radiation can also transmit energy. For example electric radiant heaters, which make use of electrically heated filaments can be used to provide heat in the rooms of houses.

(i) State the name of the type of electromagnetic radiation emitted by a radiant heater [1]

(ii) The body of a human being emits radiant heat at a rate of 450 W. Clothing reduces this heat loss by about 150 W. 200 W of energy are absorbed by the human body from the walls and floors of a room. Calculate the rate at which energy must be provided by an electric radiant heater in order to maintain the energy balance of the body. [1]

(d) The Sun has a surface temperature of approximately 6000 K and emits electromagnetic radiation with a typical frequency of 6×10^{14} Hz which corresponds to yellow light. The red star Betelgeuse has a different surface temperature to that of the Sun; it emits radiation with a typical frequency of 4×10^{14} Hz.

(i) Suggest what can be deduced about the surface temperature of Betelgeuse from the fact that Betelgeuse is a red star. [1]

(i) Calculate the wavelength of the radiation emitted by Betelgeuse. You will need to use the equation

speed = frequency × wavelength

(The speed of light = 3.0×10^8 ms^{-1})

Give your answer in standard form and with appropriate units. [3]

[Total: 13]

Unit 2 Examination questions

1 Figure 1 shows the directions of ocean currents in the southern part of the Pacific Ocean. The ocean currents are driven by the prevailing winds. The coupling of ocean and atmosphere also leads to an area of high atmospheric pressure near South America and an area of low atmospheric pressure near Australia.

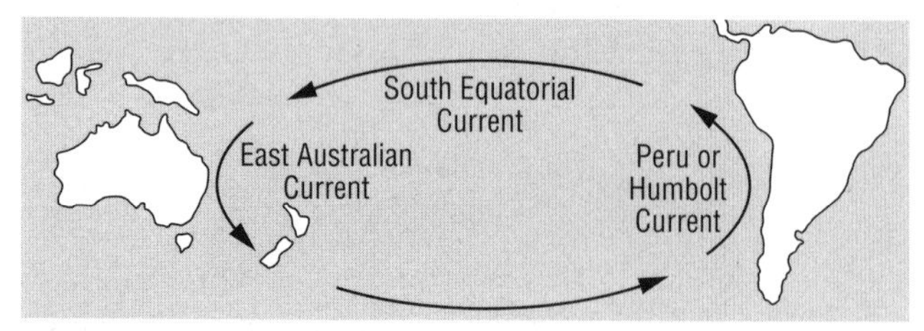

Figure 1

(a) State two factors, other than prevailing winds, that influence the direction of ocean currents. [2]

(b) At intervals of 4–10 years, this pattern of ocean/atmosphere behaviour changes. The warm equatorial current now flows from Australia to South America and the positions of the areas of low and high pressures are reversed. This is called an 'El Nino' event – it is also known as the 'southern oscillation'. Typically it lasts for about 9 months.

(i) Suggest why the term 'southern oscillation' is used to describe such an event. [2]

(ii) Suggest two ways in which an El Nino event affects the climate of the neighbouring region of South America. [2]

(c) Water has an unusually high specific heat capacity.

(i) What is the meaning of the term 'specific heat capacity'? [1]

(ii) Explain why the high specific heat capacity of water has a significant effect on the temperature of a coastal region. [2]

(iii) Hydrogen bonding is one type of intermolecular bonding present in water. How is a hydrogen bond formed? [3]

(iv) Explain, in terms of the behaviour of molecules, why water has an unusually high specific heat capacity. [3]

[Total: 15]

(OCR A-level Science 2846 Jun06)

2 The increase in carbon dioxide in the atmosphere may also have an effect on the oceans. This is because carbon dioxide behaves as an acid when it is dissolved in water. As carbon dioxide concentration rises, this may cause changes to the chemistry of sea water – for example changes in its pH which, in turn, will affect the ecosystems in the oceans.

(a) (i) Sea water has a pH of 8.2. What happens to the pH of sea water when it becomes more acidic? [1]

(ii) Give an example of an environmental effect which can occur in an aquatic ecosystem when the pH of the water changes. [1]

(b) The acid formed when carbon dioxide dissolves in water is known as carbonic acid, H_2CO_3. This can behave in the way shown in equation 1:

Equation 1 $H_2CO_3 \rightarrow H^+ + HCO_3^-$

(i) Explain how equation 1 shows that H_2CO_3 is an acid. [2]

(ii) Complete equation 2, which shows a further process that can occur:

Equation 2 $HCO_3^- \rightarrow \_\_\_\_ + CO_3^{2-}$

(c) Any pH changes caused by carbon dioxide are expected to be small. The presence of other gases in the atmosphere, such as sulfur dioxide and nitrogen oxides, has caused much larger changes in certain parts of the world, such as Scandinavian lakes.
Suggest and explain one reason why these gases produce a greater effect on the pH of water than that produced by carbon dioxide, even though these gases make up a much smaller percentage of the atmosphere. [2]

(d) (i) State one way in which the effects of acidic pollutant gases, such as those in part (c), have been minimised. [1]

(ii) Do you think this would be an effective way of preventing the acidification of the oceans by carbon dioxide? Explain you answer. [1]

[Total: 8]

(OCR AS Science 2842 Jan07)

3 **(a)** Deoxyribonucleic acid (DNA) is made from a long chain of nucleotides, each of which consists of a deoxyribose sugar, a phosphate group and a base (adenine, guanine, cytosine or thymine). Draw a diagram to show how these components are arranged in a short section of double-stranded DNA. [4]

(b) Name the type of bond that holds the two strands of DNA together. [1]

(c) Explain how just four different bases in DNA are sufficient to code for all the different amino acids in proteins. [3]

(d) Ribonucleic acid (RNA) is another molecule composed of nucleotides. In RNA, the sugar is ribose instead of deoxyribose.

(i) State two ways, other than the type of sugar, in which RNA is different from DNA. [2]

(ii) Two different kinds of RNA are found in a cell – transfer RNA (tRNA) and messenger RNA (mRNA). Describe how both are involved in the synthesis of proteins. [4]

[Total: 14]

(OCR A-level Science 2844 Jun06)

4 The temperature at the centre of a star can be hundreds of millions of degrees. The energy needed to maintain this temperature is provided by nuclear fusion reactions. The equation for one such reaction is shown in equation 3:

Equation 3 $^{3}_{1}H + ^{2}_{1}H \rightarrow ^{4}_{2}He + ^{1}_{0}n$

(a) What is the meaning of the term 'nuclear fission'. [2]

(b) Write down the numbers of protons and neutrons on the left-hand side of the nuclear reaction shown in Equation 3. [2]

(c) Draw a labelled diagram to illustrate a simple model of the structure of an atom such as helium. Use the terms proton, neutron and electron in your diagram. [3]

(d) Scientists have not been able to investigate atoms by viewing them directly. Instead, evidence about the structure of atoms has been gathered from indirect observation. Describe briefly one piece of indirect evidence which has contributed to our knowledge of atomic structure. [3]

(e) Equation 4 shows an equation for a nuclear fission reaction that occurs in a nuclear power station:

Equation 4 $^{1}_{0}n + ^{235}_{92}U \rightarrow ^{90}_{38}Sr + ^{144}_{54}Xe + 2^{1}_{0}n$

Using the information in Equations 3 and 4, describe two differences between nuclear fusion and nuclear fission. [4]

[Total: 14]

5 Ozone (O_3) plays an important role in the atmosphere by absorbing ultra-violet radiation from the Sun. However since the early 1980s there have been concerns over the effects on human health resulting from a reduction in the ozone concentration in the atmosphere.

(a) **(i)** Name the layer of the atmosphere in which most ozone is found. [1]

(ii) Describe how a reduction in the ozone concentration in this layer can affect the health of humans at the surface of the Earth. [2]

(b) It is thought that the loss of ozone is the result of the release of CFCs (chlorofluorocarbons) into the atmosphere. These break down to produce chlorine atoms in the atmosphere. The manufacture and use of these chemicals has now been banned and replacement chemicals identified.

(i) State one large scale use of CFCs in the past [1]

(ii) Complete equation 4 to show the breakdown of a CFC.

Equation 4: $CF_3CCl_2F \rightarrow$ _______ + Cl [1]

(iii) State two factors which you would need to take into account in selecting a suitable replacement for a banned CFC. [2]

(c) The breakdown of ozone occurs in a two stage process in which the Cl atom acts as a catalyst to dramatically increase the rate of ozone breakdown.

(i) Complete equations 5 and 6 to show how this process occurs:

Equation 5: $O_3 + Cl \rightarrow ClO +$ _____

Equation 6: $ClO +$ ____ $\rightarrow Cl + O_2$ [2]

(ii) All the substances in this process are gases. Is the chlorine acting as a homogeneous or a heterogeneous catalyst in the breakdown of ozone? Justify your answer. [1]

(iii) Explain how a catalyst increases the rate of a chemical reaction, such as the breakdown of ozone. [3]

[Total: 13]

Answers

Examination answers

Unit 1

1 (a) A community of living organisms [1] and the physical environment in which they live [1]

(b) (i) tree → insect → frog/bird/snake [1]

(ii) Several organisms may feed on a single type of organism lower in the food chain. [1]

(c) $\frac{310}{2200} \times 100$ [1] = 14.1% [1]

(d) The average temperature/sunlight intensity is greater in a rainforest [1] the average rainfall is greater in a rainforest [1]

[Total: 6]

2 (a) (i) The wavelength is shorter in visible light [1]

(ii) They both travel at the same speed/they both consist of vibrating electric and magnetic fields [1]

(iii) Near infrared [1]

(b) Absorption [1] and (re-)emission [1]

(c) (i) Visible light has been reflected from the land, thermal infrared has been emitted/visible light shows the reflectivity of the land, thermal infrared shows the temperature [1]

(ii) The (tops of) the clouds are cold [1] so emit very little infrared radiation [1]

(iii) White [1]

[Total: 9]

3 (a) Nitrates/ammonia/ammonium compounds/urea [1] and phosphates [1]

(b) (i) Increased rate of growth of plants [1] the extra nutrients cause plant growth [1]

(ii) Plants eventually die [1] due to competition for e.g. sunlight [1]. The plants are decomposed by bacteria [1] This uses up oxygen (which is necessary for fish to survive) [any 3 points]

(c) (i) Run-off/dissolve in rain-water draining into waterways [1]

(ii) Use organic farming methods/calculate most efficient level of fertiliser use/introduce barriers to prevent run-off [1]

[Total: 9]

4 (a) (i) Absorb [1]

(ii) Diffract [1]

(iii) Refract [1]

(iv) Reflect [1]

(b) Cones are sensitive to narrow ranges of wavelength/only one colour [1] this produces colour vision [1] rods are sensitive to the whole visible range, producing black and white vision [1]

(c) (i) Infrared [1]

(ii) 100 W [1]

(d) (i) Belelgeuse has a lower surface temperature [1]

(ii) 7.5×10^{-7} [1]; answer given in standard form [1] unit = m [1]

[Total: 13]

Unit 2

1 (a) Rotation of the Earth/Coriolis effect [1] Deflection by land masses [1]

(b) (i) Occurs in a repeating pattern [1] in the southern hemisphere [1]

(ii) Climate is drier [1] and warmer [1]

(c) (i) Energy required to heat 1 g/1 kg of water up by 1 °C/1 K [1]

(ii) Water can store a large amount of energy [1] it can transfer heat from high to low latitudes, creating a warmer climate [1] OR Water takes a long time to heat up or cool down [1] temperature of coastal regions is cooler in summer/warmer in winter [1]

(iii) A δ^+ H atom [1] is attracted to a δ^- atom/oxygen [1] which is small/has a lone pair [1]

(iv) A lot of energy is required to make molecules vibrate [1] because the force between the molecules is strong [1] this force must be overcome when molecules vibrate [1]

[Total: 15]

2 (a) (i) Becomes lower/closer to 7/just below 7 [1]
(ii) Populations of aquatic animals change/corals dissolve/shells of shellfish dissolve NOT just 'fish die' [1]
(b) (i) It releases H^+ ions [1]
(ii) $HCO_3^- \rightarrow H^+ + CO_3^{2-}$ [1]
(c) The acids produced by nitrogen and sulfur oxides/nitric and sulfuric acid are strong acids [1] they ionise completely/produce more H^+ ions
AW the release of nitrogen and sulfur oxides produces a relatively large change in the % of these gases in the atmosphere [1] release of CO_2 causes only small changes in the % of CO_2 in the atmosphere [2]
(d) (i) Scrubbers in chimneys/catalytic converters on car exhausts/desulfurisation of fuels/liming of lakes [1]
(ii) Depends on answer to **(i)** – answer is likely to be NO, reasons include: difficult to capture carbon in same way as NO_x and SO_x/captured CO_2 may be dumped in sea anyhow/cannot remove C from fuels/cannot lime seas as they are too large etc. [1]
[Total: 7]

3 (a) Sugar bonds to phosphate [1] in an alternating sugar–phosphate backbone [1] base is bonded to sugar [1] two bases from different strands bond together [1]
(b) Hydrogen bonds
(c) Bases arranged in codons/sets of 3 [1] there are 64 ways of arranging four bases [1] enough combinations to code for all (20) amino acids [1]
(d) (i) RNA contains uracil, not thymine [1] RNA is single stranded [1]
(ii) mRNA carries information about the base sequence/a copy of the genetic information [1] from the nucleus to the ribosomes [1] tRNA recognises a codon (on the mRNA) [1] carries a (specific) amino acid to the ribosome [1]
[Total: 19]

4 (a) 2 nuclei/atoms join together [1] to form one larger, heavier nucleus/atom [1]
(b) 2 protons [1] 3 neutrons [1]
(c) Central nucleus [1] labelled with words 'proton' and 'neutron' [1] electron shown in shell/orbit round nucleus
(d) Description of experiment [1] results of experiment [1] deduction about atomic structure [1]
e.g. alpha particles fired at thin layer of metal [1] some particles deflected through large angles [1] (positive) nucleus is present
OR alpha particles were collided with (nitrogen) atoms [1] hydrogen nuclei were formed [1] the nucleus contains protons [1]
(e) Similarities [2]: new element produced; neutrons produced; energy released; two particles collide
Differences [2]: two atoms join in fusion, one atom splits in fission; fission can cause a chain reaction, fusion cannot; fission normally occurs for heavy nuclei, fusion normally occurs with lighter nuclei; fusion requires higher temperatures than fission
[Total: 18]

5 (a) (i) Stratosphere [1]
(ii) More ultra-violet reaches the Earth's surface [1] causes skin cancer/cataracts/sun-burn [1]
(b) (i) Propellants (in aerosols)/refrigerant fluids/solvents/blowing agents for foams [1]
(ii) CF_3CClF/C_2F_4Cl [1]
(iii) Boiling point/ozone depletion potential/toxicity/flammability/stability in atmosphere, etc. (any two) [1]
(c) (i) Equation 5: O_2 [1] Equation 6: O [1]
(ii) (Homogeneous) catalyst and reacting molecules are in the same state [1]
(iii) Provides a new mechanism/pathway [1] with a lower activation energy [1] more collisions are successful AW [1] [1]
[Total: 11]

(AW = alternative way of answering a question)

Spread answers

Answers to activities in the spreads

Some of the spreads are in the form of a long activity, based around a series of questions.

Answers to these questions are given below:

1.1.7 Analysing remote sensed images (I)

1 Band I: 4×10^{-7} m – 1.1×10^{-6} m, Band 2: 5.7×10^{-6} m – 7.1×10^{-6} m, Band 3: 1.05×10^{-5} m – 1.25×10^{-5} m

2 Band 1: mostly visible light (+ some near infrared), Band 2: near infrared, Band 3: thermal infrared

3 Clouds appear white in the reverse image

4

Image A		
Feature	**Appearance**	**Reflectivity**
Cloud	White	High
Land	*Grey*	*Medium*
Sea	*Black*	*Low*

Image B		
Feature	**Appearance**	**Relative temperature**
Cloud	*Black*	*Low*
Land	White or pale grey	High
Sea	*Dark grey*	*Medium*

5 Ice crystals

6 (a) Top surface of clouds is very cold, so emit very little infrared
(b) Ice or water droplets reflect (or scatter) light very well
(c) These clouds are very tall so the top of the cloud is very high in the atmosphere (and hence very cold)

7 (a) N. Africa is mostly desert, containing pale sand which reflects light very well
(b) The sand is very hot (during the day) and is a good emitter of thermal infrared

8 Figure 1a would be almost completely black (except for the lights of cities). Figure 1b would be similar, although the land would probably appear a little darker (because it will be cooler at night)

9 Titan is covered with dense clouds so visible light reflected from the surface cannot reach sensors

10 Suggests the surface of Titan is rough and mountainous. However some areas (black in the image) may be flat and smooth

1.1.8 Analysing remote sensed images (II)

1 Band 1: Green, Band 2: Red, Band 3: Near infrared

2

	Appearance	Reflectivity
Green light (a)	Dark grey	Low
Red light (b)	*Dark grey*	*Low*
Near infrared	*White*	*High*

3

Feature	Reflection of Band 1	Reflection of Band 2	Reflection of Band 3	Appearance in image
Sediment	High	Low	Low	Blue
Clear water	Low	Low	Low	Black
Snow	High	High	High	White
Vegetation	Low	Low	High	Red
Bare rock	Low	Low	Low	Black/grey

5 Dark blue areas may be rivers or bodies of water containing sediment, light blue areas may be drier deposits of sediment e.g. on flood plains of rivers

2.2.3 Clearly a problem

1 Algae population has changed – free-floating has been replaced by filamentous on lake bed; fish population has fallen because fish fail to breed and mature fish died

2 Acidic pollutant gases have made the lakes acidic; acidification occurs naturally as the lake becomes more mature; land use around the lake has changed

3 Sulfur dioxide, nitrogen dioxide

4 Sulfur dioxide comes from the combustion of sulfur compounds which are impurities in coal and oil; nitrogen dioxide formed when nitrogen and oxygen from the air react in the high temperatures of car and lorry engines

5 Indirect evidence

6 Diatoms

7 Pollen grains (providing evidence about land use)

2.4.13 Analysing epidemiological data

1 Ratio of controls: cases should be 14.2 to 1

2 Actual ratios group 1 = 6.6 to 1, group 2 = 17.6 to 1, group 3 = 5.7 to 1, group 4 = 16.6 to 1

3 Group 2 and group 4 have ratios similar to the theoretically expected value

4 Group 1 = 2.1 times greater, group 3 = 2.5 times greater
5 The difference between the observed and theoretical values is quite large which would suggest it is statistically significant; however in the groups 1 and 3 the number of cases is fairly small. So the difference could be due to chance
6 People living close to power lines might be poorer than people living further away
7 The observer might have been biased in the way they recorded the readings
8 Because cancer leukaemia cases are relatively rare. It is very difficult and expensive to carry out such large scale studies

Answers to numerical questions

1.1.1

2 (a) 2.31×10^5
(b) 1.298×10^3 m

1.1.2

2 (a) (i) 9.7×10^3
(ii) 1.69×10^{-3}
(iii) 6.82×10^9
(iv) 4.92×10^{-9}
(b) (i) 2.89×10^4
(ii) 1.7×10^{-2}
(iii) 2.089×10^6
(iv) 3.89×10^{-5}
3 (a) 1.5 m

1.1.4

3 20 times greater

1.1.5

3 0.158 m

1.1.9

5 (a) 6.4×10^7 m^2
(b) 8000 m

1.2.3

3 (a) 8.5×10^5
(b) 50%
(c) 12 000 J
(d) 30 500 J transferred as heat; 71%

1.2.8

3 (c) 4.0 kJ ha^{-1} yr^{-1} passed on to animals, 26.2 kJ ha^{-1} yr^{-1} returned to soil from decay of plants, 2.5 kJ ha^{-1} yr^{-1} output from ecosystem

2.1.2

1 (a) 5 Nm^{-2}
(b) 1000 Nm^{-2}
(c) 10 Nm^{-2}
2 (a) 16 000 N
(b) 0.00016 N
(c) 1600 N
3 15 dm^3
4 5 dm^3

2.1.7

3 (a) 1350 kg m^{-3}
(b) 0.00074 m^3

2.2.4

4 0.287 g dm^{-3}

2.2.8

2 (a) 3.3×10^{13} Hz
(b) (i) 9.0×10^{-4} cm
(ii) 1111 cm^{-1}

2.4.1

4 15 708 kJ kg^{-1}

2.4.4

3 (a) (iii) 3.5×10^{-25} kg
(b) (ii) 7.7×10^{-29} kg

2.4.9

2 (a) 48C
(b) (i) 4 J
(ii) 4 W

2.4.10

2 600 Ω
3 400 A
4 (a) 1.76×10^8 W
(b) (i) 1.76×10^6
(ii) 1%

Practice answers

Unit 1 Module 1

1 (a) (i) Radio waves, infrared, visible, ultraviolet, gamma
(ii) Gamma is most damaging as it carries the most energy
(b) Diffraction, reflection, refraction

2 (a) (i) Waveband: electromagnetic radiation with a range of wavelengths / frequencies
(ii) A small square in an image (corresponding to a fixed area of land)
(b) (i) Clouds consist of ice crystals and water droplets which reflect / scatter visible light
(ii) Tops of clouds are cold

3 (a) The colour displayed in the image is not the same as that of the radiation it represents
(b) (i) Red
(ii) White

4 (a) (i) 2.00×10^8 Hz
(ii) 1.5 m
(iii) Probably not – wavelength is much smaller than the gap between building

5 (a) White light can be thought of as consisting of red, green and blue light. Chlorophyll absorbs red and blue light. The remaining wavelengths of light (green) are reflected
(b) Green light is absorbed by the green cones in the retina. These produce an electrical signal, which our brain interprets as indicating the colour green
(c) Moonlight has a much lower intensity than sunlight, which is too low to stimulate the cones. The rods detect this low level light but they detect all wavelengths of visible light so the brain only interprets signals from the rods as a 'black and white' image

Unit 1 Module 2

1

Chlorophyll	Absorbs light energy and transfers it to chemical energy
ATP	Product of respiration; acts as a mobile energy store
Glucose	Product of photosynthesis; acts as a store of chemical energy
Carbon dioxide	Used as the source of carbon atoms in the manufacture of biomass by photosynthesis
Lipid	Component of the cell membrane

2 (a) (i) Trees are being cut down, destroying habitat; sections of rainforest are being isolated from each other
(ii) Overfishing is removing a complete species from the ecosystem
(b) (i) The variety of organisms present in an ecosystem (number of differnt species and variation within species)
(ii) Rainforest plants provide many medicines; rainforest plants are the wild versions of many crop plants and the genes may be needed to improve our modern versions of these crops

3 (a) leaves → ants → anteaters → jaguars
(b) Most energy is lost in respiration or decay (of leaves)

4 Original population had some variation; some individuals were better adapted to compete for food, mates, etc.; these individuals survived and passed on the advantageous characteristic; if populations were isolated in different environments then different characteristics would develop; eventually new species would be produced

5 (a) Glucose stores energy in the form of chemical energy; ATP is a mobile form of this chemical energy
(b) (i) 93 moles of ATP
(ii) 32% of energy is transferred to ATP

6 (a) (i) A process which opposes or reverses a change which occurs in a system
(ii) Addition of fertiliser increases amount of nutrients in soil; growth of plants is encouraged by these extra nutrients; growing plants absorb nutrients from the soil; nutrient level returns to its original state
(b) (i) 400–500 °C, 100 atm pressure, iron catalyst
(ii) Requires fossil fuel to be burnt to provide energy to maintain the high temperature and pressure; fossil fuels release pollutants, e.g. CO_2
(iii) Overuse of fertilisers can cause eutrophication; nitrates running off into waterways cause rapid plant growth which eventually causes low oxygen levels

Unit 2 Module 1

1 (a) Molecules of gas are in random motion. They collide with the walls of a container and exert a force as they bounce off the walls. Pressure = force on a fixed area (e.g. 1 m^2)

(b) Diagram should show high pressure at poles and 30°N and S, low pressure at equator and 60°N and S

2 (a) High boiling/melting point, high specific heat capacity, density of solid is greater than density of liquid

(b) (i) Lone or non-bonding pair of electrons

(ii) Single covalent bond

(iii) Small positive charge

3 (a) From SW to NE; wind blows from high to low pressure; but is deflected clockwise by rotation of the Earth / Coriolis effect

4 (a) Off the S coast of Greenland

(b) Dense, salty water is cooled by wind; makes it more dense than surrounding water and so it sinks

5 (a) Ice melts, revealing bare rock below; this absorbs sunlight better than ice; less sunlight is reflected so Earth heats up more rapidly

(b) Plants may grow faster if temperature and CO_2 levels increase; they will carry out photosynthesis more rapidly which absorbs CO_2; this may reduce CO_2 levels

6 (a) CO_2 released by burning biofuels is absorbed while biofuel crop is growing

(b) Large scale growing of biofuels would cause shortage of land for food crops / would require large scale used of fertilisers which require energy for manufacture

Unit 2 Module 2

1 (a) (i) Coal contains sulfur compounds as impurities; these form SO_2 by reacting with oxygen when coal burns

(ii) Between 1 and 6

2 (a)

	Where found	Type of radiation	Trend in concentration
Ozone	*Stratosphere*	Ultraviolet	Decreasing
Carbon dioxide	Troposphere	*Infrared*	*Increasing*

(b) CFCs

(c) Propellants in aerosols, refrigerant fluid, solvent for cleaning, blowing agent for foams

3 (a) $N_2 + O_2 \rightarrow 2NO$; $2NO + O_2 \rightarrow 2NO_2$

(b) (i) +2

(ii) +4

(c) oxidation of N atoms increase

(d) (i) $O_3 + O \rightarrow 2O_2$

(ii) Provides a new pathway for the reaction; this may have a lower activation energy

4 (a) 3.4×10^{-6} m

(b) Wavelength = 3.4×10^{-4} cm; $1/3.4 \times 10^{-4} = 2941$

(c) Rest of spectrum has a high % transmission except for a sharp drop (='peak') at 3000 cm^{-1}

Unit 2 Module 3

1 (a) (i) They catalyse the important reactions which occur in cells
(ii) Component of cell membranes, structural role in e.g. muscle or hair, some hormones and antibodies are protein molecules
(b) (i) This is where the substrate molecule bonds and reacts
(ii) Increasing temperature can cause change in shape; changes in pH, presence of inhibitors
(iii) Changed shape means that the substrate cannot fit and/or bind so the activity of the enzyme is reduced

2 (a) Crop could be made more able to survive drought conditions (e.g. gene for longer roots introduced); gene for production of pesticide could be introduced so it will not be eaten by pests (other examples also)
(b) Virus, inert particle (e.g. gold), plasmids from bacteria

3 (a) (i) GCGAATGCC
(ii) 3
(iii) Sequence of three bases determines which amino acid is incorporated into the protein

4 (a) (i) Small 3-dimensional features such as sheet or helix
(ii) The overall complex 3-dimensional structure of the protein
(b) (i) Peptide links (covalent bonds)
(ii) Hydrogen bonds

5 (a) The sequence of bases present in an organism
(b) Introducing a gene for antibiotic resistance into a bacteria; expose the bacteria to the antibiotic – if it survives then the gene transfer has been successful
(c) Many possible answers: e.g. superweeds – gene for herbicide resistance passes to wild plants which may then grow uncontrollably

Unit 2 Module 4

1 (a) Emission of CO_2 causing global warming, emission of SO_x causing acid deposition, emission of NO_x causing photochemical smog (and acid deposition)
(b) (i) 23%
(ii) Suitable pie chart with angles as follows: coal 83°, oil 144°, natural gas 83°, other 50°
(iii) Nuclear, hydroelectric, tidal, wind energy etc.

2 (a) 400 000 V
(b) Reduces power loss by heating
(c) (Step-up) transformer

3 (a) (i) 15 960 J
(ii) 14 280 J
(b) Ethanol 31 920 J g^{-1}, hexane 57 120 J g^{-1}
(c) Hexane, more energy released from same mass

4 (a) Atoms with the same number of protons but different numbers of neutrons
(b) ${}^{14}_{6}C \rightarrow {}^{14}_{7}N + {}^{0}_{-1}e$
(c) β particles are ionising, so they can cause damage to cells e.g. cause cancer

5 (a) (i) Region of space in which a magnetic pole / current carrying wire experiences a force
(ii) Direction of current alternates, so direction of induced field will also alternate
(b) (i) 1636 A
(ii) Energy = power × time; time = 21 600 s; energy = 360 × 106 × 21 600 = 7.78 × 1012 J
(iii) Power lines suspended at a great height; field strength falls with distance; cables arranged so that fields will cancel out; e.g. by placing cables with currents running in opposite directions close to each other

Index

absorption 10, 11, 20
acid–base titration 79
acid rain 75, 76, 79, 80–1
acids 72–3
  neutralisation 78–9
  as pollutants 73, 75, 76–7, 80
activation energy 83, 96
active site 96–7
active transport 29
adaptations 36
adenine 98, 99
ADP 27, 29, 30
aerobic respiration 30
air masses 48, 49, 52
air movement *see* atmospheric circulation
algae 72, 76
alkalis 79
alpha decay 122
alpha particles 117, 120, 121
alternating current 128, 133, 134
alternating field 132–3, 134–5
alternative energy *see* renewable energy
aluminium ions 80
amino acid sequence 102, 103
amino acids 94, 95, 97
ammonia 42, 55
ammonium nitrate 42
anaerobic respiration 30
analogue signals 12
Antarctic ice 66
antibiotic resistance 104, 107
anticodons 102, 103
aperture 7, 12
atmosphere 20, 48
  changing conditions in the 51
  water vapour in 89
  see also acid rain; greenhouse gases; ozone
atmospheric circulation 48–9, 52–3
atmospheric pressure 48, 49, 52, 53
atomic number 118, 119, 122
atoms 54, 116–17, 118
  in chemical reactions 74
  electronegativity 56
  emissions from radioactive atoms 120–1
  oxidation state 78
  see also electrons; nucleus
ATP 27, 29, 30
autotrophs 26, 27, 28
  see also plants

bacteria 27, 38, 39
bacterial cell 26
balancing equations 74, 75, 115
barcodes 13
base pairs 98, 99, 100, 101
behavioural adaptations 36
beta decay 122
beta particles 120, 121
billiard ball model 116
biodiversity 34–5, 36–7
biofuels 65, 113, 127
biological hazards 120–1, 124–5
biomass 26, 32, 115
biomes 32
biosynthesis 26
'blind watchmaker' model 36
blue 9, 11, 15, 19, 20
Bohr, Niels 118
bonds
  bond angles 54, 55, 57
  breaking and forming 82–3, 112–13
  ionic bonds 95
  polar bonds 56, 88
  see also covalent bonding; hydrogen bonds
Boyle's law 50–1, 51
Bt corn 105
burette 79
butterflies 36–7

calcium carbonate (limestone) 78, 80
calorimetry 112
cameras 7, 12
cancer 13, 82, 120
  leukaemia 135, 136
  treatment 35, 121
car emissions 75, 76, 81, 83
carbon capture 65
carbon dioxide 26, 27, 30, 86
  concentrations 66, 88
  emissions 114
  and global warming 33, 34, 62
carbon isotopes 62, 119
carbonate ions 78
carbonic acid 72
case–control studies 135, 136
catalysts 42, 82–3, 84, 85
catalytic converters 81, 83
cell membrane 28–9
cells 26
  enzymes and proteins roles 95, 96
  passive and active transport 28–9
CFCs 84–5, 88
characteristics 36
charge-coupled devices 21
Charles' law 51
chemautotrophs 27
chemical energy 26, 27, 28
chemical equations 5, 74, 75, 115
  nuclear 122
chemical reactions 74
Chernobyl 125
China 65, 81, 115
chlorine 84, 85
chlorophyll 10, 11, 26, 27
chloroplasts 31
clay soil 80
clean technology 81
clean-up technology 81
climate
  analysing climate data 66–7
  dry and wet 49
  and ecosystem productivity 33
  and oceans 58
climate change 62–5
  and evolution 37
  positive feedback and 41
  see also global warming
climate zones 49
clouds 15, 17, 49, 53, 89
coal, combustion of 113, 114, 115
codons 102
cohort studies 135
cold fusion 127
colour 8–9, 10
  primary colours 9, 19
  pseudocolour and false colour 15, 19
  see also blue; green; red
community 32
complementary base-pairing 99, 100, 101
concentration 73, 79
  enzyme activity and 97
  greenhouse gases in the troposphere 88
concentration gradient 28, 29
condensation 53, 57
cones 9, 21
consumers 32
contamination 121
convection 53
Coriolis effect 52, 58
covalent bonding 54–5, 56
  vibration 86–7, 88
crops 40, 104–7, 106
crude oil, competition for 124
cumulative effect 36
currents
  electrical 128, 130, 132–3, 134
  oceanic 58–9, 61
cysteine 95
cytoplasm 30, 31
cytosine 98, 99

Darwin's finches 36
decomposers 32, 39, 43
denitrifying bacteria 39
density 56
  energy density 113
  flux density 132
  sea water 60–1

ibose 98
lfurisation of fuels 81
toms 77
iffraction 6, 7
diffraction grating 8
digital signals 12
direct current 130
disulfide bridges 95
DNA 98–9, 100–1, 102, 104, 105
donor organism 104
dot-and-cross diagrams 85
double helix 98, 99, 100
droughts, predicting 67

Earth
  infrared radiation from 11, 88
  remote sensing of 14–15
  rotation of 52, 58
  temperature of 10, 11, 49
  see also atmosphere; climate; climate change; global warming
ecology 32, 34
ecosystems 32–3
  nutrient cycles in 38–9, 40
  rainforests as 34, 35
  water-based 39, 42–3
El Niño event 58, 59
electrical circuits 130–1
electrical current 128, 130, 132–3, 134
electrical field 4, 132–3
electricity 128–31
electrocution 131
electromagnetic field 133
electromagnetic radiation 4, 5, 13
  see also gamma radiation; infrared radiation; solar radiation; ultraviolet radiation
electromagnetic spectrum 5, 14
electron-pair repulsion theory 54
electronegativity 56
electrons 54–5, 118
electrostatic forces 54
elements 38
emission spectrum 20
end point 79
endoplasmic reticulum 100
energy
  increasing cost of 124
  law of conservation 31
  released by fuels 112–13
  stored energy 30, 31
  see also activation energy; chemical energy; kinetic energy; light energy; renewable energy; solar energy
energy density 113
energy efficiency 65
energy-flow diagrams 31
energy generation, future of 124–7
energy profiles 83
energy reserves 124
energy transfer 10, 28, 30–1
environmental effects
  energy generation and 124–5
  fossil fuels 114–15
  global warming 64
environmental monitoring 14–15
enzymes 96–7, 100, 104
epidemiology 135, 136–7
equations *see* chemical equations
ethanol 87, 113
eutrophication 39, 42–3
evaporation 53, 57, 89
evolution 36–7
eye 9, 21
eyesight 8

false colour images 15, 19
feedback 33, 38, 40–1, 64, 89
fertilisers 39, 40, 42–3
field line 132
fields 4, 127, 132–5
first law of thermodynamics 31
fish 35, 43, 72, 76
Fleischmann, Martin 127
flue gas desulfurisation 81
fluid mosaic 28
flux density 132
food chains 32
food webs 32, 35
force 50
  electrostatic forces 54
  intermolecular forces 56
  maintaining protein structure 95
fossil fuels 114–15
  combustion 76, 112–13, 114
  desulfurisation 81
  reserves of 124
frequency 4, 5, 7, 8, 12
  rods and cones 21
  thermal radiation 11
  vibrating bonds 86–7
fuels, comparing 113
  see also biofuels; fossil fuels

gamma radiation 5, 13, 120, 121
gases, pressure and 50–1
genes 36, 98
  marker genes 104, 105, 107
genetic code 102
genetic engineering 104–5
genetically modified crops 104–7
genome 104
geographical isolation 36, 37
global warming 33, 34, 41, 86–7
  and greenhouse gases 62–3
  predictions 64
  sinking and 61
  see also climate change; greenhouse effect; greenhouse gases
glucose 26, 27, 29, 30
golden rice 105
green 9, 10, 15, 18–19, 21
greenhouse effect 64, 86–7
  see also global warming
greenhouse factor 88
greenhouse gases 34, 62–3, 88–9
  models and 64
  strategies for limiting 65
  see also carbon dioxide; methane
grey-scale images 15
guanine 98, 99
Gulf Stream 59, 60, 61

Haber process 42
habitat 32, 34–5
Hadley cell 52, 53
half-lives 125
hazards *see* safety
HCFCs 85
Helsinki protocol 1985 81
herbicides 105, 107
heterogeneous catalysts 83
high pressure 48, 49, 52, 53
hockey stick graph 63
homogeneous catalysts 83
horizontal air movement 52
human health *see* safety
humans
  climate change and 65
  energy balance of 11
humus 80
hurricanes 53, 59
hydrocarbons 114, 115
hydrochloric acid 72
hydrogen, as fuel source 113, 127
hydrogen bonds 56–7, 95, 99, 100
hydroxide ions 78

ice 56, 57, 61, 66
ice ages 63, 66, 77
ice-core sampling 63, 66
ideal gas laws 50–1
images 12
  thermal 15, 16–17
  see also satellite images
indicators 73, 79
induced electrical field 133
information transfer 12–13, 101, 102
infrared radiation 5
  from the Earth 11, 88
  global warming and 86–7
  near infrared 14, 15, 18
infrared spectroscopy 86–7
inhibitors 97
inorganic compounds 26
inputs, nitrogen cycle 38, 39, 40
insecticides 106
inter-tropical convergence zone 53, 67
interglacial periods 63
intermolecular forces 56